BETWEEN THE ROCKS AND THE STARS

Between the Rocks and the Stars

Narratives in Natural History

Stephen Daubert

Illustrations by Chris Daubert

[VANDERBILT UNIVERSITY PRESS * NASHVILLE]

This book is printed on acid-free paper.
Manufactured in the United States of America

Library of Congress Cataloging-in-Publication Data

Names: Daubert, Stephen, author.
Title: Between the rocks and the stars : narratives in natural history /
Stephen Daubert ; illustrations by Chris Daubert.
Description: Nashville : Vanderbilt University Press, [2020] | Includes
bibliographical references and index. | Summary: "Presents a collection
of vignettes from the wild, each of which describes the natural
advantage of a particular organism. These true-to-life accounts are then
posed in particular circumstances that illustrate the
principles-commensalism, speciation-that shape the place of these
organisms in their living environment"—Provided by publisher.
Identifiers: LCCN 2019030998 (print) | LCCN 2019030999 (ebook) | ISBN
9780826522740 (hardcover) | ISBN 9780826522757 (paperback) | ISBN
9780826522764 (ebook)
Subjects: LCSH: Natural history.
Classification: LCC QH45.2 .D375 2020 (print) | LCC QH45.2 (ebook) | DDC
508—dc23
LC record available at https://lccn.loc.gov/2019030998
LC ebook record available at https://lccn.loc.gov/2019030999

COVER IMAGES: Butterflies and flower, illustration from *Insects of Surinam*,
1726 (Natural History Museum, London, UK/Bridgeman Images); Tachycineta
bicolor, Tree Swallow, from John James Audubon's *Birds of America*, 1827–1830
(Natural History Museum, London, UK/Bridgeman Images); Red bandfish,
Cepola macrophthalma from Edward Donovan's *Natural History of British
Fishes*, 1802–1808 (© Florilegius/Bridgeman Images); Precious stones and crys-
tals, Chromolithography of Mineralogy by Gustav Adolph Kengott, 1886
(© Florilegius/Bridgeman Images)

[CONTENTS]

[PREFACE]

WHEN OBSERVANT TRAVELERS in the natural world find something new, their first impulse is to ask "What is that?" They want to know more about what they have encountered. The discipline of natural history contains that knowledge—the sum of our experiences with the natural world. It provides the information that enriches the moment of discovery for these observers.

The breadth of the discipline of natural history knows no bounds. Its subject matter is all-encompassing—it includes the living things and the nonliving environment that supports them. Its province goes back through the fossil and geological records—there is no limit to how far back. It reaches out into the infinity of the cosmos from which the Earth and sun formed long ago. It includes all we have learned about Earth's organisms. It is organized around universal principles that apply everywhere. It provides the context within which our appreciation of the natural world reaches its highest levels.

Our experiences with the natural world take the form of encounters—each of them transpiring in a single instant in time. But within its natural history context, each of those instants expands. A single creature we have discovered becomes just one branch point on a family tree populated by scores of its evolutionary relatives. A second network—of competitors, predators, and parasites—branches across the first tree. All the organisms on Earth are found to be connected to all the others through networks of interaction and relatedness. An understanding of that matrix of connections is part of the discipline of natural history.

Through the prism of our understanding of their deep natural context, our understanding of each organism extends back through lineages of its ancestors and predecessors. Our excursion into the ecology of organisms describes the habitats in which they prosper. Those habitats eventually reach marginal boundaries in which the organisms can no longer exist. Our understanding of these life histories may track changes in habitats over the eons and chronicle the accommodations made by the inhabitants as they adapt to the changes. Still farther back, we may meet the most primordial of ancestors at the dawn of life. They arose after our planet came into being, condensed out of star dust in deep space.

The chapters in this book each begin by introducing one particular facet of the natural world—a creature, a landscape. Some of those introductions occur down in the rocks at ground level, others out in the forest, or out at sea. Others take place on larger scales, reaching back in time or out into the galaxy.

These vignettes are not embellished. (Citations that document the bases of the descriptions are provided in the Notes sections that follow each chapter.) Events are presented to appear just as they would in encounters anyone could observe—should that person happen to be there, at the right place and the right time. (However, events in many of these chapters take place quite a long way off the beaten path.)

After the introductory encounter, these chapters proceed to put their particular events into the contexts of their underlying natural histories. That context reveals what came before, what other advents may rise to challenge the status quo, and what we stand to learn from the situation. Why would laurel and magnolia flowers be pollinated by beetles instead of bees? Why would sunfish stare at the sun? How do the microbes of summer that live in flower nectar survive a flowerless winter? How do fireweed seeds under the ground in springtime know that a wildfire passed above in the fall? These questions are discussed in Chapters 3, 9, 12 and 17. Many other natural world encounters are set in their natural history contexts elsewhere, *Between the Rocks and the Stars.*

— I —

Ant Butterflies

THE FLOOR OF the rainforest is a silent space, despite the diversity of life it sustains. The creatures living there are well concealed indeed. They have been getting better at deception over quite a long time. And their predators—the smaller dinosaurs, which have evolved into the modern birds—have been keeping pace. They have been getting better at seeing through the deception. They have ways of finding the hidden inhabitants that live there on the ground in front of them.

The low-living, ground-dwelling invertebrates are cryptic, clothed in the colors of mold and fallen leaves. They don't draw attention to themselves by moving. But on rare occasions a susurration arises from the undergrowth. It transforms the flat spaces into a sea of commotion. An advancing tide hisses through the deadfall on hundreds of thousands of tiny feet.

The opportunistic birds in the branches notice every sound that disturbs the early morning stillness. They are attuned to all the signs of life in their realm, and they were listening for this particular disquiet. They glide down for a closer inspection, converging from every direction. Their feathered friends from farther afield follow the calls of the earlier birds.

The rustling in the leaves is a raid of army ants, attacking everything in its path. The horde behaves as a single being—each individual connected with all the others. They organize their formations with no central commander. They pass pheromone signals back and forth through the air or through deposits they leave on the ground. Their advance covers every surface, flowing through the leaves, down into burrows, and up into trees—a boiling flood that stings anything it touches.

These ants have given up the subterranean ant-farm lifestyle. They are nomadic, moving from one camp to the next. They live in the open, in tents woven from their own linked bodies. The advance of their raid is quiet, a ripple in the leaf-mold accented by the occasional stroke of a leaping insect falling back to Earth. But the commotion turns raucous when the birds arrive. They are hailing their mates and calling out their pecking-order foes.

This specific guild of forest birds moves with the ants. These birds have also given up the home-based life. They no longer defend specific territories in the forest—they defend moving territories. They are wanderers, following the campaigns of the wandering ants. They are the antbirds.

Hundreds of antbird species have diversified into all the forest niches that sustain army ants. These birds make up less than ten percent of the tropical avifauna species. But their numbers can make up more than a quarter of the birdlife in areas where raiding ants march. Each of these antbirds is a specialist; many of them live solely by following the ants.

They travel one hop ahead of the advancing army. They have mastered the skill of wading through the tide without being bitten. In constant motion, they flit away just before the swarm engulfs their perch—just before they feel their feet crawling. They stand aside and peer down from vertical surfaces or from dangling vines. They see everything in fine detail, from just inches away.

The antbirds do not eat ants. They eat what the ants eat. The ants flush their prey from every hiding place. The birds pursue the spiders, scorpions, earwigs, centipedes, and roaches that crawl, jump, and run from the advancing ant army. Crickets that spring into the air may never come down—there are antbird species that launch from the sidelines to intercept them mid-arc. Beetles that fly straight up will fall prey to other kinds of antbirds, which pursue targets farther above the fray.

The bigger birds, like the ocellated antbird, command the space at the front of the raid. Smaller species, such as the bicolored antbird or the rufous-collared antbird, take the flanks. Still smaller, the scale-backed antbird hangs back and watches to hawk insects that escape into the air.

The woodcreepers perch on the wider trunks, while horizontal perches accommodate birds like ant tanagers or ant wrens. The largest species—ant pittas, or ground cuckoos—walk through the foliage ahead of the ant columns. At any time there may be scores of birds from a dozen species accompanying a large raid. They call out to each other and squabble over possession of the best overlooks.

The antbirds take advantage of the misfortune that has befallen their prey. They are raptors of disaster, like the rain birds that track the water's rising edge to hunt creatures flooded from concealment, or the firebirds scouting the advance of the ground fire that chases their prey from dry grass.

The antbirds steal much of the pillage that would otherwise be taken by the army ants. The ants themselves are blind, unaware of the theft going on all around them. When the grub they are holding is lifted off the ground and shaken until they fall off, all they can do is run faster, until they find their scent trail again. The birds can rob the army of most of its plunder.

AFTER THE FRONT lines of raiders have passed, the commotion recedes into the distance and the forest settles back down. The calling of the birds fades; the vacated foreground lies still. A narrow, quiet rivulet of ants runs both ways through the center of the glade, carrying nondescript fragments of raiding booty back to the ant bivouac.

The musty odor of army ants lingers in the air. It attracts those forest dwellers attuned to that particular airborne scent: the ant butterflies. Gliding in silence, these sprites materialize in the aftermath of the raid. Some of them float ghost-like on invisible, transparent wings. Their flight is lazy and slow; they frequently dip down to ground level. But they are not pursued by the antbirds that bring up the rear of the raid. The birds know that these butterflies are distasteful. Many of them share the red-and-black warning coloration worn by the long-wings—butterflies that feed on toxic passion vines.

Ant butterflies are of different sizes and color patterns. Each species has its own ways of avoiding its predators and parasites. The larvae of each feed on different host plants. But all the ones that have showed up here have one thing in common: they are all female. The males of the species are off elsewhere in the forest.

Another thing they share is a niche—one of the many hyper-specialized niches in the tropics. Like the antbirds, these insects know how to detect and follow ants. But what they are really following is the antbirds.

The swath of ground passed over by the army ant raid is speckled with white dots, dropped along the route by the birds. The butterflies touch down on these droppings, which are sources of essential, high-quality nutrients. The quiet flyers attend close to the rear of the raiders, to sup from the white spots while they are still moist.

The ant butterflies can assimilate nitrogenous compounds excreted by the birds, for use in the production of their own eggs. Other butterflies lay a relatively small number of eggs; they are limited by the amounts of nutrients they were able to accumulate in their bodies as larvae. But adult ant butterflies can add to the amount they hatched with. They are longer-lived than others of the lepidoptera. They extend their egg-laying success across their longer life—sustained by what they can find to forage.

THE ACTIVITY IN this green world is a zero-sum game, where losses in one place are balanced by gains elsewhere. Nothing in the forest is wasted. The addition of butterfly eggs to the foliage, and the addition of the antbirds' eggs to their own hidden nests, balances the ruination wrought by the army ants. Alternating cycles of destruction and growth sustain the diversity of life here—maintaining the vitality concealed beneath the trees.

The ant butterflies gather where the ants have gathered. They fly in from far and wide to the place where they will find antbirds. These butterflies spend their lives in pursuit of a movable feast, the most concentrated source of butterfly nutrients to be found in the forest. Then, when the feast is over, they glide higher into the forest canopy. Like the departing souls of the creatures that died below, they rise in the calm after the catastrophe has moved on, returning peace to the clearing.

NOTES

Eciton burchelli, in the American tropics, is the archetypical army ant legionnaire (Franks et al., 1991). Their communication with each other and with their environment is olfactory (by pheromones and scents) and tactile (by touch). A soldier's eye consists of only a single optical facet. It tells them the difference between night and day and not much more. The operations they launch are diurnal and predictable, well suited to kleptoparasitism by the antbirds (Willis & Oniki, 1978). There are more than a hundred species of army ants. *E. burchelli* is exceptional in the broad net it casts for prey.

Many of the other species are specific predators on the hives of other ants or social insects.

Many different butterflies follow army-ant raids, including skippers (Lamas et al., 1993) and members of the *Ithomiidae,* long-winged mimics of the Heliconiid butterflies (Ray & Andrews, 1980). This guild of understory butterflies follows the birds that follow the ants. In the first published description of the ant butterflies, Drummond (1972) did not notice the connection between these butterflies and the high-nitrogen sustenance they derived from the birds. Their dependence on antbird droppings has since been documented, e.g., by Ray & Andrews (1980), who noticed the preponderance of females among them. The ithomiids carry dissuasive alkaloids in scent pockets in their clear wings. Those toxins protect them from predatory birds, and from spiders—which cut them out of their webs (Brown, 1984).

The antbirds are raptors of disaster, as are the firebirds, such as the black kites. That kite, from the Australian outback, has been reported to rekindle the fires that flush their prey from the tall grass; they have been seen carrying twigs that are burning on their far ends (Montague, 1970). This raises the possibility that once upon a time, at the dawn of human history, these raptors, by their example, taught another two-legged animal about the handling of fire.

REFERENCES

Brown. K. S. 1984 Adult-obtained pyrrolizidine alkaloids defend ithomiine butterflies against a spider predator. *Nature* 309, 707–9.

Drummond, B. A. 1972. Butterflies associated with an army ant swarm raid in Honduras. *Journal of the Lepidopterists Society* 30, 237–38.

Franks, N. R., et al. 1991. The blind leading the blind in army ant raid patterns: Testing a model of self-organization. *Journal of Insect Behavior* 4, 583–607.

Lamas, G., et al. 1993 The Ahrenholz technique for attracting tropical skippers (*Hesperidae*). *Journal of the Lepidopterists' Society* 47, 80–82.

Montague, A. 1970. A remarkable case of a tool-using bird. *American Anthropologist* 72, 610–14.

Ray, T. S., & C. C. Andrews. 1980. Ant butterflies: Butterflies that follow army ants to feed on antbird droppings. *Science* 210, 1147–48.

Willis, E. O., & Y. Oniki. 1978. Birds and army ants. *Annual Review of Ecology and Systematics* 9, 243–63.

Whale Hill

CASCADE CANYON DESCENDS from the mountains between silent walls, on a morning long past, by an ancient South American coastline. On this day, the stream that cut its gorge is just a trickle sliding over polished stone. It flows over a short season, sometimes remaining dry for years. The sandy bottom of the canyon flattens out and spreads across a beach, beside steep, shaded cliffs on a desert shore. When it does flow, the stream follows a shallow channel across the sand—its bed braided with dark streaks of pyrite. The flow slows as it passes through the shadows of the cliffs; rivulets disappear into the sand before they reach the sea.

Cool air now hangs above the isolated shore. Seabirds have deserted their towers overlooking the water. The beach is still except for the cadence of the surf. But the shoreline stands transformed. A powerful El Niño storm has pummeled the area in the previous days. High winds and rain assaulted the landscape. The beach is littered with driftwood and seaweed.

During the storm, wind-driven rains saturated talus slopes across the mountains. Boulders slipped down landslide slopes and splashed into the runoff. Minerals were leached from the rocks and the dust. Streams carrying dissolved iron and phosphate accelerated through the watershed and into the canyon. For hours afterward, Cascade Creek ran at flood stage. The beach suffered one of its once-in-a-hundred-years transformations.

The tame little stream above the coast swelled into a cataract. It descended with enough force to dig up the canyon's sandy floor and carry it into the ocean. The muddy torrent plowed a trench across the beach. It deposited tons of sand in shoals offshore. The stream course from the canyon mouth to the breakers was transformed into a sea-level lagoon twenty feet wide and just as deep. The new channel crossed the shoreline from the breakers to the cliffs and extended up into the mouth of the canyon. A rocky notch that formed the canyon's true base, usually buried in sand, became exposed. When the sediment settled out, the rocky crevasse at the base of the cascade showed clearly through twenty feet of still water.

NOW, IN THE aftermath of the storm, swells riding in from the sea continue all the way across the beach. They travel on the surface of the new lagoon until they reflect off the base of the vertical wall behind. That diversion then sends them up the canyon as ripples that pass above two parallel rows of pillars.

The pillars have been have exposed on the bottom. They stand like fence posts rising from the mud-stone flooring, extending halfway to the surface. Between them, they enclose a narrow space twenty feet long. The posts in the two rows curve inward toward each other as they rise, tapered at their ends into blunt points.

These are the ribs of a giant cetacean—a partially exhumed whale skeleton lying on its back. The sand that covered it centuries earlier is now stripped away. From above, the ribs appear to sway back and forth with the passing ripples of dappled light.

The last of the storm's suspended minerals trickle down Cascade Creek, enter the pool at the canyon mouth, and swirl past the rows of pillars. The fresh water floats along in a layer atop the warmer salt water in the lagoon, until the two are mixed by the breakers. Then the runoff enters the sparkling South Pacific and spreads away on the tide toward the far western horizon.

IN CONTRAST TO the desolate coastal strand, life flourished off this shore. Beds of kelp and eelgrass spawned a food chain that rose through tight schools of fingerling sardines and anchovies all the way up to the largest predators. Near in, the birds cried above the water. The heads of seals and the fins of sharks tracked across the surface. Thousands of square miles of ocean stretched further away, animated by flocks of petrels and gannets. Distant misty geysers of whale spout hung in the air all around the compass. The rising and falling phrases of whale calls carried from the depths to echo up through the surface, haunting the quiet beach.

In that long lost epoch, the waters off of Cascade Beach supported a diversity of sea life unimaginable in today's era of ocean exploitation. The community of creatures that lived there was supremely productive and resilient. But the changes brought by the El Niño storm would test that resilience.

A day after the storm, the subtle scent of the sea had already become sharper. An edge of foam was building up on the wrack of kelp along the strandline. Minerals that had been rare in the water—limiting the growth of marine microbes—were now being delivered in excess in the runoff from the mountains. More nutrients also appeared in the deep water, where storm-driven currents were welling up from the depths. With these new supplements, the marine microbes began increasing their numbers exponentially.

By the second day the water was turning. The faces of the breakers showed a telltale opacity—the sea surface itself was off-color. The populations of microbes were growing into competition with each other. Some of the microscopic dinoflagellate algae responded to the competition by fluorescing. They gave off tiny scintillae of light when disturbed. Where they were numerous in the water, the small animals that preyed upon them found themselves outlined in glowing flashes when they moved. This revealed their positions to their own predators—to the benefit of the dino-flagellates they preyed upon. The larger consequence was phosphorescence of the surf. As night fell, a shimmering shore-break lit the arc of Cascade Cove in electric blue from one end to the other.

BELOW THE WAVES, tiny diatoms were engaged in chemical warfare. They excreted a mix of compounds into the water that poisoned the other microbes that competed with them for nutrients and for sunlight. The toxins increased with the growing numbers of diatoms, disabling the cope-pods that preyed upon them. Other microbes produced compounds that subdued the krill that preyed upon them. These battles for survival would have collateral effects on the bystanders in their ocean.

The third day after the storm, the sea took on tints of color. Creamy patches offshore marked blooms of cyanobacteria. They graded to purple nearer the beach, where pink clouds of diatoms mixed with the other microbes. In the depths, the defensive molecules excreted by the microbes had reached toxic levels.

The turmoil below soon emerged above the surface. Movement in the surf resolved into the forms of large animals slogging through the breakers. They waded closer to the shore, exposing themselves—they looked like polar bears, but with longer snouts and long, curved forelegs. These were marine sloths, strangers to the land, but now abandoning their ocean. They were clumsy in the shore-break—unable to support themselves out of water, struggling just to breathe.

More of them appeared farther out. Some of those were no more than lumps passively rolling, awash in the swells. The lead animals hauled themselves up the beach, stopping to rest in the waves, disoriented, eyes wide. They lay on the sand, like boulders covered in sea-moss, exhausted, buffeted, unable to progress further except when a wave pushed them on up the slope.

A pod of gray dolphins sped through shore-break, easily avoiding the struggling sloths. Their momentum carried them well up onto the strand. Their flippers slapped the wet ground, and they arched back and forth, but for all the commotion they showed no interest in finding their way back into the water. Eventually they tired and just lay in the sun, their backs drying for the first time in their lives.

Farther down the cove a silvery blue marlin lay still on its side along the strandline. In the other direction a more rounded form had appeared. The waves rolled it over, revealing a single tusk that ran the length of its body—a walrus whale. All manner of marine creatures were coming ashore.

As the day evolved, the beach came to squirm with animals abandoning the water. They had absorbed the poisons from the algal bloom or eaten crustaceans or shellfish tainted with microbial toxins. After a while their struggles subsided. The shore grew littered with their stranded bodies.

Overnight, those dead creatures began to bloat. In this condition, they became buoyant. So as the tide rose, they came to life once more. They advanced further up the beach, moving a few feet at a time with the strongest waves. The surf rose farthest up the shore where the storm had deepened the stream channel. Riding the surge, the invasion of the dead floated through the new lagoon, toward Cascade Canyon.

The drifting herd of bodies tumbled through the canyon mouth and spread out across the pool of water inside, filling the base of the canyon. The following morning, the first of the condors arrived to find them on the edges of the pool between the rock walls.

THAT NEXT DAY would see a major addition to the animal assembly on the beach. Far offshore, shapes appeared on the surface, sailing on the sea breeze that blew into the cove. They took hours to move in close enough for the scavengers gliding over the cliffs to recognize them as upside-down whales. They had died at sea and bloated. Now they floated on their backs, their bodies mostly out of the water, white bellies bright in the sun. Each one sailed head-first, flukes trailing like a sea anchor. None of them carried recent gashes in their fins or flanks from the megalodon sharks; those giant scavengers, like the rest of the sea life, had been laid low by the poisoned ocean.

The breeze came up across the shore in the afternoon to whistle through the canyon gap. It drew the floating whales along with it into the cove. Their great size became apparent as they approached the surf-line and ground to a halt on the beach. The waves broke against them, rising only halfway up their flanks. Those breakers animated the distended bodies, jostling and aligning them, nudging them farther up the shore as the tide came in.

The wind peaked at sunset, along with a high spring tide. Waves came across the beach through the lagoon. A group of the huge, passive whales floated toward the cliffs, into the temporary channel from the canyon. Moving a few feet at a time with the highest of the swells, they eclipsed the entire canyon mouth as they washed against the rocks. At the urging of the surge from behind, the pod of dead beasts advanced single file, aligned by the wind as they sailed toward the base of Cascade Creek. Their advancing bulk filled most of the surface of the temporary pool.

CONDORS DESCENDED ON the feast that had washed up on the coast. They tore into the dead mammals on the strand line and quickly ate their fill. Then they spent their newly acquired energy bickering with each other. They had no ill effects from eating the tainted meat. This was their niche—they were immune to the sea-borne microbial toxins, which would cause paralysis and organ failure in most other animals.

In the afternoon, they took to the air. A soft on-shore breeze hit the cliffs and was deflected upward, buoyed by thermals rising off the sand. The birds spiraled effortlessly thousands of feet into the sky. From on high, they could scan fifty miles of coastline. Dark lumps up and down the white strand showed the extent of the die-off in the ocean. On the land side, the barren piedmonts had come alive with a magenta wash of flowers after the storm.

The higher the condors orbited the beach, the farther away they themselves could be seen. The vortex of black wings low on the horizon was noticed by other sharp-eyed high-fliers patrolling other beaches to the north and south. Those other condors were drawn from fifty miles away to investigate the mass stranding in this cove. More arrived every day. Eventually all of the condors on the Pacific Coast of South America would glide in on the updrafts along the cliffs to visit this site, most of them staying for weeks. Land scavengers, which could rip dead animals apart and chew through the bones for the marrow, were barred from the shoreline by the steep desert mountain range. The condors had this feast all to themselves.

ON THE OTHER side of the breakers, the status quo ante was returning as the bloom of diatoms died out. Viruses that attack the bacteria and algae proliferated—each virus infected a single microscopic organism, multiplying its own numbers a thousand-fold in the process. The viral numbers grew while the microbial population crashed.

Populations of bacteria that could metabolize toxin molecules appeared offshore one day and disappeared the next. The dinoflagellates' ranks were thinned by predators that could resist their toxins—rotifers, amoebae—and those predators bloomed. Krill appeared to prey on them in their turn, and schools of herring were drawn in from the open ocean to feed on the krill. The water became blue again. And the sand became white again. The condors skeletonized the carrion on the beaches and let the tides scatter the bones.

Their largest feast lay piled inside the canyon, but for much of the time it remained out of reach, under water. The condors could only look down and wait. One by one, the whales floating on the lagoon exploded. The birds roosting on the cliffs flinched in the dark as each blast echoed between the steep walls. The exploded whales sank, and the whales floating behind them moved up with the swells to take their places toward the head of the lagoon. Eventually, the whales all stacked themselves on the bottom at the base of the streambed. The only carrion available to the winged scavengers lay half-submerged along the base of the canyon wall.

The gasses of putrefaction escaped from the sunken carcasses. The lagoon bubbled noisily with the emerging hydrogen sulfide. Hydrogen sulfide is heavier than air, which it displaced, filling the low spaces. An invisible pool of the toxic gas spread out in the canyon above the lagoon. The pool grew deeper when the sea breeze blowing across the canyon mouth kept the gas from flowing out across the beach.

Condors are not repelled by such an odor, but they learn to avoid high levels of it. The white-ruffed adults lined the rim above and waited for the wind to change. They sat in the sun and looked down in silence for days.

They watched the occasional all-black juvenile condor glide right past them and dive into the valley of death below. The young bird would alight and stand for a moment on the mass of collapsed bodies on the shore; then it would immediately take off again. Unfortunately, condors cannot fly straight up. The juvenile would attempt to rise by flying along the length of the canyon, but that entire length was filled with an invisible pool of poison gas.

Soon the flapping condor would lose strength and crash, cartwheeling into the sand. The adult birds watched this scene play out over and over. Sometimes the naive juveniles would bank off to the side to escape the suffocation, only to collide with the wall. Such birds died in seconds. The adults' keen eyes could count the black-feathered remains of all of them— piled or scattered across the canyon or sunk in the lagoon.

Eventually their hunger drove the birds to give up on the tons of inaccessible meat and seek sustenance elsewhere. They faced into the wind, opened their wings, and stepped off into the sky. Floating out through the mouth of Cascade Canyon, they drifted away above the clean white strand and rose to disappear to the north or south.

UNDER THE CONSTANT urging of the onshore current, the piles of sand deposited offshore at the mouth of the lagoon began to migrate back to the beach. The waves worked to smooth out the bottom, pushing sand back up the slope. The sand went farthest where it met no resistance, at the entrance to the lagoon. The temporary channel began to fill. It filled more quickly at high tide during days of wind-driven storm surge. The in-filling sand displaced the brackish water, burying the animals on the bottom and returning the beach to its natural contour. The whale graveyard disappeared into the ground.

Seawater evaporated from the shallow lagoon where the rising slope of the sand refilled the canyon. Over time, the intermittent trickle of Cascade Creek displaced the sea salt from the water table under the mouth of the canyon. Below the surface, the buried creatures decomposed, leaving only skeletons. Sulfides produced by dissolving soft tissue reacted with dissolved iron carried in the groundwater to precipitate pyrite on the bones. Small voids in the bone were slowly filled with dissolved silicates and carbonates. Buried sharks, made only of cartilage, disappeared from the assemblage, leaving only their teeth. The hollow bones of the birds shattered as the entire assemblage gradually compacted and settled.

Though rare, the conditions that brought these dead animals to rest in Cascade Canyon would occasionally recur. The tide would run red once again, and a mass die-off would litter the shore with creatures of all kinds. Storm surf would ride the high tide upland, and more bones would be deposited, new layers above the old. Some would again be buried with minimal disturbance, and the fossil bed in the base of the canyon would grow deeper.

As the millennia passed, the level of the flat sand surface rose meter after meter with the rise in sea level, and with the geological descent of the coastal margin. The weight of the over-burden grew above the buried grave-yard. When the burial level reached one hundred feet, the pressure became great enough to squeeze the water out of the mineralized skeletons, fossil-izing them into rock in a sandstone matrix. The transformation progressed so gradually that the bones retained their shapes down to the finest details.

Millions of years finally reconfigured the shoreline geometry, erasing the canyon and the cove where the whales had come to rest in the shelter of rock walls. The cliffs eroded; the beach moved farther out and then came back in. The sea level rose and fell by hundreds of feet. Where the ground had slumped, sediment piled up on the sea bottom, burying the fossilized creatures deeper. Later the area was uplifted by tectonic forces, raising it once again above sea level.

As our era approached, the tectonic rise of the Andes Mountains tilted the terrain, uplifting the ground and bringing the fossil bedrock upland, to where some of the bones were exposed by erosion.

Those geological changes reshaped the entire western shore of the island of South America. A volcanic land bridge rose to connect its northwestern edge to the continent to its north. The Paleoindians crossed that bridge

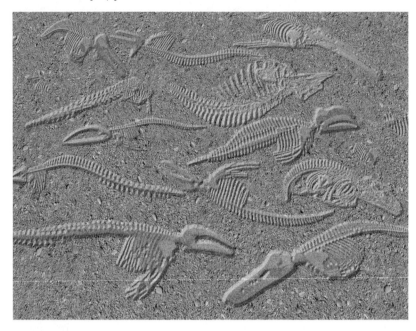

and eventually came to the Chilean shore where this story had unfolded. They saw the exposed bones on the hillside and named it for its fossils. A thousand years later Spanish conquistadors renamed the area Cerro Ballena (Whale Hill).

Five hundred years after that, a road crew dug into the fossil bed and exposed its full extent. Paleontologists from the Smithsonian Institution and from local universities were able to document the scene. They recorded all the details that allow us to imagine those long-lost days—an unprecedented snapshot of the vitality of that prehistoric shoreline. Then they stood aside while the road crew pressed ahead. The fossil bed was returned to the subterranean condition it had known for the previous millions of years—buried once again, under the northbound lanes of the new Pan American Highway.

NOTES

The events depicted here are based on descriptions by Nick Pyenson and his colleagues at the Smithsonian Institution of Washington, DC, and their co-workers from other institutions, in the *Proceedings of the Royal Society*

(2014). They documented a collection of prehistoric animals from South Pacific waters that was somehow assembled and buried without disturbance on an ancient beach. The Pleistocene shore of South America was home to a cast of marine mammals that included the Walrus Whale (*Odobenocetops*) and the marine sloth (*Thalassocnus*). In this scenario an extensive toxic bloom of marine algae (e.g., McCabe et al., 2016) poisoned the marine fauna. Dead and floating marine mammals were not attacked by megalodon sharks because the die-off at sea had killed or driven off almost everything that lived in the waters, including the large marine scavengers. And the beached bodies were not disturbed by terrestrial scavengers because they came ashore onto a barren desert. The ones that came to rest in a particular sheltered depression were buried intact, only to be revealed during a road-building excavation millions of years later.

REFERENCES

McCabe, R. M., et al. 2016. An unprecedented coastal toxic algal bloom linked to anomalous oceanic conditions. *Geophysical Research Letters* 43, 10.366–10.376.

Pyenson, N. D., et al. 2014. Repeated mass strandings of Miocene marine mammals from the Atacama region of Chile point to sudden death at sea. *Proceedings of the Royal Society B* 281 (1781) 2013.3316.

— 3 —

Spicebush

THERE IS A particular sensation that enlivens a springtime morning. Down in the canyons in the Pacific Northwest's coast range mountains, its scent rises through the stillness under the trees. Not a floral bouquet—it is fruitier than that—with the tang of pineapple, perhaps, or strawberry. Some would say it has the aroma of a wine cellar. This airborne flavor rouses an air of expectation in a particular group of wild-land creatures, and they come to seek out its source.

The fragrance can arise in two ways. It is produced by the fermentation of the juices of senescing, overripe fruit. Or it can be constructed from scratch—synthesized by a newly opening flower.

In the first instance, yeast cells generate the scent as a byproduct of their growth. This happens when they move from the skin of a fallen currant or laurel berry and begin to grow in the exposed pulp. In the second case, the same chemical scent is synthesized by the flowers of the western spicebush. Either way it is produced, it attracts one particular group of creatures. They are not looking for nectar, or for fruit that has ripened fully. These animals seek the last stages—the berries that have fallen from the branches. They are guided in their search by the fruity aroma.

IN THE UNDERSTORY of those North American coast range mountains, on May Day, the scent of these flowers can fill a canyon pathway. The trail, dappled in light filtered through the branches, travels not just downhill. It travels back down through time as well—as will any trail beneath trees whose roots go way back into the realm of prehistory.

Higher up the ridgeline, the canyon supports a stand of bay laurel. The grove scents the sea-breeze with camphor and eucalyptol—scents that emanate from the laurel's bark and leaves. Farther toward the base of the ravine, the steeper walls contain a stand of coast redwoods.

Spicebush thrives in the shade beneath these trees all the way down the canyon. Its own heady bouquet of aromatic leaf scents (for which the plant is named) is similar to that of its confamilial laurels.

These plants are relics of the dinosaur ages. Their progenitors first appeared one hundred million years ago. The descending canyon glades are not alive with butterflies and bumblebees; showy, sweetly scented blooms are not common here. The insects that visit flowers for their nectar had not yet arisen back in the Cretaceous Period. Pollination then was the work of an earlier-arising insect order—the beetles. The Cretaceous plants in these canyons have brought their beetle pollinators down to the present with them.

The western spicebush, for example, is a relic from the dawn of the flowering plants. The blossom is dark against the green of the glen—a woody maroon color. It does not attract butterflies or bees. The bloom is instead pollinated by tiny sap beetles.

The flower retains an archaic structure, from the days when flower petals had not yet differentiated from the leaves that enfold the developing bud: the sepals. Spicebush petals grow directly from the tip of the branch on which the flower stands. The male and female sexual organs are both contained in the same bloom—not born on separate plants, or even separate flowers.

Northern spicebush, found on the east coast of North America, shows the evolutionary progress that has transpired from the Cretaceous through the current era. Its flowers are now showy yellow; they have come to attract hover flies and swallowtails. The sexes have separated: each flower is either male or female. Its fragrance is sweet and floral. Western spicebush is a distant relative of northern spicebush. The western lineage has not kept up the evolutionary pace—it is a relic from the past.

WHEN ITS LONG thin petals first begin to unfurl, and its scent of fermented fruit fills the glade, the western spicebush flower is mobbed by flies that don't often visit flowers: fruit flies. Female fruit flies are attracted to the scent of fermenting fruit, upon which they seek to lay their eggs. The males are attracted to the scent because they expect that female fruit flies will be there. Fruit flies mate on softening fruit, in which their larvae will grow.

But there is no fermented fruit here—the fruit flies have been misled by the spicebush flower. And the scented bud is still tightly closed at its center. The flies cannot access most of the dark blossom—they are not pollinators of the Western Spicebush.

In many perennial plants, each flower is either male or female. The sexes are separated across the distance between blooms. But in these ancient spicebush plants, both sexes are present in the same flower. This could lead to self-pollination, which would compromise the function of the flower, which is to facilitate cross-pollination between separate spicebush plants.

Yet the sexual organs in the spicebush flower are in fact separated from each other—not across distance, but across time. When the female parts of the bloom are receptive, and the scent that attracts pollinators emanates from within, the flower's male pollen has not yet been shed.

The presence of fruit flies indicates the ripeness of the female part of the spicebush flower. Another insect that feeds on over-ripe fruit is also attracted: the diminutive sap beetle. Unlike the flies, the beetle has short, compact legs held close under its body. Its wings do not hinder its walking through tight spaces: they are sheathed beneath the animal's shell.

Smaller than a fruit fly, the sap beetle has the leverage to squeeze between the folded petals and down into the heart of the flower, where the scent of ripe fruit is strongest. These beetles feed on yeasts that often cover fallen fruits. Their larvae grow in the pulp of grounded, moldering fruit.

But the sap beetles that enter the spicebush flower find no fruit pulp; no fungal ferment is spreading. All they find is a tight space filled with the plant's receptive stigmas. And to deepen the disappointment, they discover that they cannot back out of the space the way they came in.

Once a beetle enters, it is trapped inside the flower bud. Layers of fine, down-pointing bristles, like those of the insectivorous pitcher plants, allow the beetles to move down into the flower, but prevent them from climbing back out. So when the next sap beetle comes to investigate the same flower, the tight space will grow twice as crowded. Four or five sap beetles may come to share the close quarters; they may pair up and mate before they depart.

The beetles are stuck inside while the flower goes through its cycle. They do not starve, however. The plant provides starchy carpels within the closed bud, which the beetles feed upon. And if the arriving beetles carry pollen from other spicebush plants, the flower will be pollinated as they move about their crowded confines.

A few days later, the attractive scent is gone, and so are the fruit flies. The female part of the spicebush flower is no longer receptive to pollination. At that point, the pollen is shed from the male anthers, covering the captive beetles. The yellow dust does not stick to ants or to random beetles or bugs that may drop in to inspect the opening bud. But it does stick to these sap beetles.

Finally the flower petals open all the way. The edible carpels wilt, and the beetles are released to fly off. By then other spicebush flowers in the glade will have begun to release their scent, inviting the pollen-dusted insects to continue the cycle. The beetles can spend a significant part of their adult lives sheltered within the folded blooms. Eventually, the females set out to find real decomposing plant material, in which to establish their next generations.

THE WALK THROUGH the shady understory in a deep, still canyon may dislodge a rock or two to slide away from its sloping setting. A fallen log may roll over to reveal the coiled millipedes or a mouse nest that it sheltered. Examination of those microcosms will reveal another scent that accompanies a cool springtime morning: the aroma of living, freshly turned earth.

This aroma is the signature of a single specific chemical that is produced by the moldering soil on the forest floor. It is synthesized by particular soil microbes that grow as invisible filaments. They elongate through the ground, extending their branching threads into whatever they contact. The scent itself is harmless, but it serves as an indication of the toxicity of the fungal filaments. The toxins they produce render them inedible; their chemistry also arrests the growth of other microbes, including the yeasts with which the filaments compete.

After they have fed from their flowers, the beetles that pollinate spicebush must find their way to the ground to continue their generations. Their larvae feed on the yeasts that live in fallen fruit. The females find the fruit that has fallen by following the scents of fermentation. The beetles will be in competition with the fruit flies, whose larvae eat the same thing. The beetles lay their eggs in the soil, and their larvae enter grounded fruit from the bottom. The fruit flies lay their eggs on the top of the fruit.

These insects that live on fermenting fruits are very sensitive to the scents of their surroundings. Fruit flies can detect scores of different scents in their environment. The scents of fermentation lead them to their food. But if they find the scent of living, freshly turned earth on the fallen fruit they discover, they will not lay their eggs upon it. They have found it too late.

The larval stages of these yeast-eating insects will not mature in fruit in which the yeast have been killed and the pulp poisoned by the threads of filamentous soil microbes. These insects live in a particular window in time. They will avoid fruit pulp that already shows the blue patches of filamentous microbe sporulation. They fly off to continue their searches across the forest floor, looking for a food source at their particular favored stage of over-ripeness.

THE SCENT OF ripe fruit does not usually emanate from flowers. But it is found in spicebush, and it can also be encountered far to the south of the spicebush groves. That distant landscape differs markedly from the temperate mountains of the Pacific Northwest. It lies in the humid equatorial jungle, on flat water alive with amphibians and birds, and even dolphins. The plants in question and their habitat are nothing like spicebush and its habitat, except on the inside—the flower is built like a spicebush flower.

Like spicebush, this tropical flower traps its beetle pollinators overnight. Similarly, its female and male organs go through their cycles of sequential ripeness in the same bud. It is another relict species that has come down to modern times unchanged from the Cretaceous Period. It is a water lily. Its flowers rise through the waterline, then open to release fruity aromas of banana and butterscotch. The plants warm the scent molecules, evaporating them into the evening air. After the butterflies and the flies have retired for the evening, the flower attracts nocturnal scarab beetles.

Each of the broad, circular leaves of the Amazonian Victoria lily cover dozens of square feet of flat-water. The large white flowers emerge next to the leaf margin. They open partially, allowing the beetles to push their way inside. Then they close on the beetles, smothering them in the folds of white satin petals. Each flower is large enough to accommodate a score of beetles.

The insects spend their adult lives floating on the water but concealed from the predatory jacanas that forage there. Those birds walk supported by extra-wide toes that stretch across the surfaces of the lily pads. The beetles imprisoned within the floating flowers feed on starchy carpels provided inside. The plants finally release their captives the next morning, after dusting them with pollen. When the beetles taste fresh air again, they mate on the flowers, and then they move on to other newly opening blooms.

They leave yesterday's plants behind. The bloom is off the older flowers—they have lost their scent, and their brightness has faded to a dark maroon. Those blossoms close again when the beetles have gone, and they withdraw below the surface to grow their fruits.

AS WITH SPICEBUSH, the Victoria lilies and their beetle pollinators first met before the advent of birds and bees. The primordial attraction between beetles and their ancient blooms suggests an evolutionary progression in the development of plant/pollinator symbiosis. In that progression, attractive colors and the promise of sweetly scented nectar arose in the post-Cretaceous Period. The relationships with birds or bees developed after beetle pollination was already established. The floral expanses of our modern spring evolved later—derived from progenitor flowers that did not offer nectar. Instead, these progenitor species induced their pollinators to visit with the promise of solid food, shelter, and mates.

NOTES

The leaves of spicebush and bay laurel exude the scent of eucalyptol, an insecticidal terpene (Scalione, 1916) especially when bruised. The flowers of western spicebush (*Calycanthus occidentalis*) are heavily pigmented, purple-brown in color, and shaped like a slate-pencil sea urchin. They exude scents associated with wine or overripe fruit. Phenylacetaldehyde, a molecule found in ferments, has that scent. The compound is derived from phenylalanine, which is derived from proteins by proteolytic degradation. Fruit flies (*Drosophila*) carry an olfactory detection system sensitive to the detection of phenylacetaldehyde (Grosjean et al., 2011). The western spicebush flower is archaic: its rays are not separated into green bud leaves (sepals) surrounding bright petals; the flowers are composed of a single type of ray (sometimes called a tepal). The spicebush flower is dichogamous: it contains both sexes, but the windows of receptivity of the respective sexual organs are separated in time (Grant, 1950). The bloom comes ripe while still closed, with its scents diffusing out to attract the beetle pollinator (the beetles are described for Carolina allspice [*Calycanthus floridus*] by Williams et al., 2008). Plants that arose in the Mesozoic, such as the cycads, and the progenitor flowering plants in the *Lauraceae* (including the bay laurel and the spicebush) and in the *Magnoliaceae*, are beetle-pollinated. Modern insect pollinators (bees, flies, butterflies, wasps) would not evolve fully into the pollinator role until the Cenozoic Era, but the beetles, including the Nitidulids (sap beetles), were already well developed in the Mesozoic. The beetles had been abroad on the land since the Permian Period of the Paleozoic Era.

The descriptions of the beetle pollination of the spicebush flowers (Grant, 1950) parallel the descriptions of the pollination of Victoria water lilies (Prance & Arius, 1975). This water lily is also dichogamous and similarly imprisons its beetles overnight while the female stigma becomes unreceptive to pollen, and the male anthers dehisce and shed pollen for export on the trapped pollinators. That archaic Cretaceous plant/pollinator relationship (Ervik & Knudson, 2003) has since evolved to create brightly colored flowers and scents that attract the active modern pollinators, which pollinate multiple blooms per hour. Those evolved plants, typified by the northern spicebush (*Lindera benzoin*) have the sexes separated in space, e.g., on separate plants.

When over-mature fruit falls to the forest floor, the scent of its fermentation is an attractant to beetles and flies. Fruit flies can detect as many as forty-six different scents in the forest (Stensmyr et al., 2012), many of which are the fermentation products of yeast growing on fallen fruit. That fallen fruit is susceptible to colonization by filamentous soil microbes such as streptomyces bacteria and penicillium fungi. Those filaments are toxic— they kill other microbes, such as yeasts (Arndt et al., 1999), as well as insects and other animals. They produce geosmin, a compound which evokes in us the aroma of freshly turned living earth. It apparently smells different to the yeast-feeding insects, to which it is a repellant (Stensmyr et al., 2012). The story of the spicebush flower was first introduced to me by G. Ledyard Stebbins, on a walk through a canyon in the interior Coast Range Mountains of California.

REFERENCES

Arndt, C., et al. 1999. Secretion of FK506 and rapamycin by *Streptomyces* inhibits growth of competing *Saccharomyces cerevisiae*. *Microbiology* 145, 1989–2000.

Ervik, F., & J. T. Knudsen. 2003. Water lilies and scarabs: Faithful partners for 100 million years? *Biological Journal of the Linnean Society* 80, 539–43.

Grant, V. 1950. Pollination of *Calycanthus occidentalis*. *American Journal of Botany* 37, 294–97.

Grosjean, Y., et al. 2011. An olfactory receptor for food-derived odors promotes male courtship in *Drosophila*. *Nature* 478, 236–40.

Prance, G. T., & J. R. Arius. 1975. A study of the floral biology of *Victoria amazonica*. *Acta Amazonica* 5, 109–39.

Scalione, C. C. 1916. The volatile oil of *Calycanthus*. *Industrial Engineering and Chemistry* 8, 729–31.

Stensmyr, M. C., et al., 2012. A conserved dedicated olfactory circuit for detecting harmful microbes in *Drosophila*. *Cell* 151, 1345–57.

Williams, R. N., et al., 2008. *Nitidulidae* found in flowers of *Calycanthus floridus Linnaeus* in northeastern Ohio. *Entomological News* 119, 397–402.

— 4 —
Getting the Jump

THE MOCKINGBIRD HUNTED through the branches, scanning the foliage for damaged leaves. Leaf damage would be an indication of feeding by cater-pillars or grasshoppers. The bird knew them to be common in these woods. But those insects usually ate all the evidence of their presence. They stayed with each leaf until it was entirely consumed—even the midrib. Those her-bivores appeared to minimize signs that might reveal their whereabouts to hungry predators; they did not leave a tell-tale trail of half-eaten leaves.

Finally the bird came upon a flowering bush, growing up into the lowest of the tree branches. Its broad leaves were shot through with holes. From above, the holes appeared black—the color of the shaded understory show-ing through from behind. And though the holes were rather small, they were all about the same size.

When the mockingbird drew closer, she noticed one hole that did not exactly look like the others: it appeared to have a faint highlight suspended in the open space. She drew above this break in the leaf to peer down directly upon it—her bill almost touching. But when she focused on the black space, it disappeared.

Quick as a wink, the leaf had repaired itself. The rent in the surface was abruptly replaced with flat, undamaged green. She had the feeling from her peripheral vision that several others of the holes had also instantaneously transformed into thin smooth foliage.

She pulled back and glanced around, appraising the situation. No explanation presented itself. The leaf just stood there, looking less shot full of holes than she thought she remembered it had been just seconds before. As before, it showed no signs of anything edible living upon it.

Lower on the plant she noticed more leaves that bore similar sets of the small, same-sized holes. Would some of those also disappear if she drew closer? Her interest in the situation was soon displaced by her hunger. Instead of further inspection, she abandoned this particular hunt and flew back to the treetops.

A GROUP OF tiny round leaf beetles had made those holes. Their habit was to stay in one place while feeding, until they had eaten their fill—feeding long enough to leave a round hole about the size of a leaf beetle. Then they moved off and rested, until their hunger returned.

Their food plants grew in shady locations. Their shells were dull black, about the color of the shade that would show through the holes they left. They hid in plain sight, among rough simulations of their own forms—the cut-out silhouettes they themselves had carved as they fed upon the foliage.

When the closest beetle noticed the approach of the massive avian predator, it paused. It tucked its muscular back legs underneath its shell, into the jumping position. It lowered its antennae into grooves beside its eyes. Then it watched to see if the predator would approach further.

These are flea beetles—able to flee like a flea does when necessary. They are very small, but the thighs of their jumping legs are (proportionally) very large—distended with extra muscle. They have a very high ratio of jumping strength to weight. So the insects are poised to release like coiled springs should the need arise.

When the predator moved in, the beetle sprang from its leaf, moving faster than two meters per second. What had looked like a dark hole turned into solid green—an unblemished patch the size of a flea beetle, uncovered in a thousandth of a second.

In an instant the insect was tumbling away through space. The beetle was smooth and aerodynamic. Its hard, round shell functioned as a full-body crash helmet. Its ballistic trajectory was interrupted half a second after launch by collision with a hard wooden stem. The sharp impact was absorbed by the beetle's shell, with no ill effect on the animal inside. The beetle was so lightweight that the sound of the impact was a tick too faint for the predator to notice.

The beetle bounced off in a new direction. Immediately it collided with a leaf, which recoiled slightly, killing most of its speed. The beetle caromed over the edge and fell to the ground.

Another flea beetle resting next to the first one had launched itself at the same instant, at a higher angle. The arc of its flight through the shadows rose uninterrupted. A few feet from its take-off point, the insect opened its wing covers, which acted like air brakes to stop its spinning. Then it raised wings that functioned like a drogue parachute to stabilize its body with the head aiming forward. The beetle controlled its flight and banked toward the shady undergrowth.

It landed on a grounded twig in the grass. After resting motionless for a minute or two, it rose up on its front legs and lifted its antennae. Then it probed the breeze for the scent of its host plant, which would inform its choice of directions for a return flight.

OTHER LEAF BEETLES find defense against their predators through bright warning coloration, which advertises a noxious chemistry in their blood. But the flea beetles that masquerade as leaf damage rely solely on their crypsis, plus their leaping abilities, to protect themselves as adults. They might be more prevalent in the forest but for the vulnerability of their larvae.

The larval stages of these dark beetles are leaf miners. They are small, flat grubs, three times as wide as they are tall. Each one lives its larval life between the top and bottom surfaces of a single leaf.

In effect, they live in a two-dimensional world. They don't know up and down. Their choices are straight ahead, with a possibility of right or left. Gravity does not show them the direction down. They never fall; they are in constant contact on all sides with their support. They grow in the flattened tunnels they excavate by eating their way forward.

Enclosure between the top and bottom surfaces of a leaf offers the larva protection from a menagerie of predators, such as ants and yellow jackets that attack leaf grazers that live in the open. But every successful strategy gives rise to a counter strategy. The leaf-mining larvae of the flea beetles are prey for a guild of parasitic wasps.

Unlike the bigger caterpillars and grasshoppers, the leaf miners cannot conceal the evidence of their feeding by consuming their leaves entirely.

The leaves are their shelter. But after a few weeks, those leaves are marked by the traces of their tunnels. Visual hunters from above can follow the white contours of the leaf mines. Those parasitic wasps locate the leaf miners by the aroma of internal leaf damage that diffuses through the tunnel walls. They can sting and inject eggs through the leaf cuticle. The wasp larvae grow within the flea beetle larvae, eventually consuming them. Most of the flea beetle eggs laid on these leaves eventually produce a next generation not of beetles, but of small parasitic wasps.

DAYS LATER, THE mockingbird was hunting along the meadow margin. She came upon an upright flowering bush, the top leaves of which were patterned with an array of shallow, same-sized scars. This reminded her of her previous leaf beetle encounter. But the foliage here was not shot through. Only the thin top layer of the surface had been eaten away in places, leaving pale necrotic windows in the leaves.

She drew closer and inspected the damaged surface. A few of the scars were more symmetrical than the others. And they were divided down the middle by a straight trace that could be the joint between beetle wing covers—though they were colored and textured like the surrounding necrotic scars.

Peering down at them, she noticed the slow retraction of oversized back legs into the jumping position beneath those wing covers. They were another kind of flea beetle, masquerading as another kind of leaf-damage. She stood back, raised her gaze to the open spaces above, then jumped from her perch to search for other prey. The sudden stroke of her wings and her kick off the branch set dozens of unseen flea beetles shooting through the air in all directions.

NOTES

The flea beetles are known for their hops. (Fleas have a similarly high strength-to-weight ratio; they also disappear by hopping away.) The flea beetles are not boldly patterned, but single-colored—the colors of leaf damage (e.g., the black of shadow, or the brown of leaf necrosis). The adults practice a form of crypsis in which they mimic the leaf damage they cause (Konstantinov et al., 2018). They masquerade as regions of destruction—features that have little value to predators, so are ignored. How they control the size and shape of the damage patterns they make is unknown. These beetles carry no denticles or spines (which would produce drag to slow their acceleration away through the air from danger). The thigh segments of their hind legs are distended with muscle, which drives their jumping (Nadien & Betz, 2016). Ballistic flight with wings closed enables a high-speed escape (up to 3.6 meters per second—Brackenbury & Wang, 1995), during which they tumble at more than fifty rotations per second. The larval stages are leaf miners in some flea beetles, root miners in others.

REFERENCES

Brackenbury, J., & R. Wang. 1995. Ballistics and visual targeting in flea-beetles. *Journal of Experimental Biology* 198, 1931–42.

Konstantinov, A., et al. 2018. Hiding in plain sight: Leaf beetles use feeding damage as a masquerade decoy. *Biological Journal of the Linnean Society* 123, 311–20.

Nadien, K., & O. Betz. 2016. Jumping mechanism and performance in beetles 1: Flea beetles. *Journal of Experimental Biology* 219, 2015–27.

One Canyon after Another

THE SHALLOW SEAFLOOR was unstable. It sloped away ever more steeply into a dark submarine canyon. The shifting silt was littered with finned scallops—fan-shaped mollusks with short wings of shell projecting from the hinge point between their two valves. The waters above them teemed with thousands of tiny larval scallops that lived among the free-swimming plankton. The scallops that had settled to the bottom were small, but farther down the fall line some of the adults were a foot in diameter.

These creatures rode an apex of molluscan diversity, here at the end of the Cretaceous Period. The shelled fauna had evolved to fill a multitude of niches by then. Filter-feeding brachiopods rose all across the seafloor like clams standing up on their hinges. Stinging cone snails patrolled the rocks, preying on fish. The sun blinked on and off, as seen from below, when schools of big-eyed nautilids moved across its light. They held their spiral shells vertical as they coasted through the water—their tentacles trailing out behind.

The scallops reacted to threats they sensed by clapping their shells together and leaping up off the sea floor. One morning, a shockwave passed through the water, and they did just that. They rose up en masse and fluttered along jerky, random flight-paths a few feet off the bottom. But as they swam, a cloud of silt from below billowed up to engulf them. The scallops fell back down and disappeared into the roiling murk. The click of their shells tumbling against one another was muted by the low rumble of a moving slope.

When the landslide subsided, the soft mud had slid a hundred and fifty feet downslope and pooled in the bottom of the submarine ravine. As the suspended sediment settled, a new canyon floor was revealed. It stood thirty feet higher than before. Its surface was smooth and level, and cleaned of everything that had lived upon it.

The scallops' second defensive strategy was to simply clamp their shells closed. Now they sat tight and waited. A few minutes after the sliding had stopped, they tentatively relaxed their adductors, but their valves did not crack open—they were pressed together from the outside. The creatures continued their efforts, motivated by the depletion of their internal oxygen. They were unaware that they were buried under twenty feet of sand and debris. Their muscles fatigued and faltered, and they suffocated.

Forty years later, an avalanche from the other side raised the canyon floor even higher. Mud flows continued through the centuries, until the hills and valleys of the sea floor in the area evened out.

Where they were deepest, the silt-flows settled of their own weight. Compression from above squeezed out the water. Buried grains of sand and clay compacted closer together. Cavities took shape in the hardening layer where it pressed around buried organic debris.

Mineralized groundwater carrying dissolved silicates flowed through the layers. As the organic matter decayed and dissolved, its molecules were replaced, one by one, by silicate molecules. Silicate crystals filled the spaces in a process so gradual that it preserved the shapes of those organic inclusions in fine detail.

Compression heating drove away more water, changing calcium carbonate to calcium oxide cement, binding the layers together. The pressure of only a few hundred feet of over-burden was enough to begin to harden the lower strata into stone.

ABOVE THE BASEMENT of sedimentary strata the biosphere proceeded with its evolution. The free-swimming ammonite cephalopods were driven to extinction during an environmental catastrophe. The numbers of molluscan fossils in the rocks above the Cretaceous boundary layer declined thereafter.

The sedimentary seabed rocks were carried along on the edge of a tectonic plate a thousand miles wide. Finally they were lifted up above sea level, rising into the sky to form rows of ridges at the collision boundary with another plate. The entire area drifted northward, into a region where the hillsides knew summer drought.

The land plants adapted to the shifting terrain. Grasses emerged as the dominant flora across summer-dry ranges. Grass-covered slopes spotted with annual flowers appeared below folded shale cliffs that marked plate boundaries. The bleached and dried stems of the grasses brightened the landscapes of autumn. Lightning-sparked grass fires returned nutrients to the soil. The fire cycle changed the mix of the broad-leaved plants all across the temperate regions.

The steepest faces of the uplifted cliffs grew unstable. Stream runoff erosion continually undercut canyon walls from underneath. Rainwater that had infiltrated into the boundary lines between sedimentary layers expanded when it froze. In winter, the most precarious spires fractured along those lines. Slabs of rock fell away into space. They fractured into a shower of smaller pieces when they crashed onto the basements hundreds of feet below.

During storms the water level rose in the streams running through the canyons. Cloudbursts over the ridges could send the water far up the banks, swelling the current even further. Flash floods raised the water pressure enough to slide even the biggest rocks in the streambeds downhill, a few feet at a time. These cascades could set boulders in motion for a second or two before they lodged in new positions. They shifted once in a decade or less, but over the millennia they tumbled down the whole length of their canyon. Their sharper edges were blunted as they repeatedly crunched against one another.

Over time, a series of landslides collapsed the canyons into the ever-deepening streams at their centers. The cliff walls were reduced to gently sloping hillsides. Fallen boulders rolled through the lengths of the streams and were deposited into the slow-moving rivers on the plains below. Where the rolling rocks came to rest, the spaces between them filled with silt, and placer deposits grew to fill the riverbeds.

With the infill, the rivers grew broader and more shallow. When the land eventually rose and the water ran off, the sedimentary placer deposits were left high and dry. Then the uplifted surface tilted, and new creeks arose to cut across the old filled riverbeds. As the new streams cut deeper channels, their banks again grew steeper. One by one the buried river stones embedded in the stream-banks fell out to splash into the water, where they waited for flash floods to wash them away in another direction. As the millennia rolled past, this downstream transport left the rocks more and more rounded.

ONE OF THESE rolled stones now lies high and dry under a bend in the bank of a seasonal stream. The stone had been left there decades earlier, at the high-water mark of a powerful deluge on a winter's night. Over the millions of years, this stone has passed through the streambeds of dozens of canyons, only the last of which now exists. When it had first fallen from its cliff, chipped out by the ice, it had been bigger and jagged. Now it was small—only a foot in diameter—and round as an egg.

Above this stone, summer-dry chaparral rose from the dusty, boulder-filled dry wash to the high ridges on both sides of a steep canyon. The scent of sage and chamise hung in the air. Shadows beneath the streamside oaks and willows shortened as the sun climbed higher. But the light was growing soft under a gathering overcast. Sky-blue faded to white and dimmed to gray. By noon, the sky looked like it does on Venus—the sun a ruddy disk behind an uneven brown pall. A wildfire was burning up the outside of the ridge.

The breeze stiffened in the afternoon, and the fire crested the upper rim of the canyon. Its volcanic glow was soon hidden by smoke that flooded downslope and obscured everything. Birds twittered to each other and took wing. Snakes deserted the grass and, side by side with their smaller rodent prey, they all moved out into the dry streambed to look for shelter in the rocks. The fire-wind strengthened, and weightless motes of glowing ash flew down the watercourse. Behind the wind, the sound of a crackling wall of flame was growing.

At its fastest, the fire moved through the grass at the speed of the wind. Its touch stole the moisture from the leaves. It consumed all the oxygen in the air, so the glowing embers entrained in the fire-wind did not burn out. In this way, the fire carried itself over the barren rocks of the streambed. It stayed alive in air from which the oxygen had been consumed. Then when the smoke thinned into fresh air on the far side, the embers ignited again, bursting into flame to rekindle the conflagration.

Flames laddered up the tree trunks from the understory foliage; their crowns wilted and then blazed to light. Standing dead snags grew into pillars of fire, then toppled down onto the wash and continued burning on the rocks.

The leading edge of the firestorm screamed through the canyon at 1,000 degrees. The round stone resting in the dry creek bed brightened beneath incandescent smoke. By weight this stone was less than one percent water; that water was bound up in solid links to the silicate matrix. But when the temperature of the surface of the rock passed the boiling point, the water broke its bonds and became mobile. Within the solid sandstone, the water flashed to steam—increasing its volume a thousand-fold, and pressing against its internal confines.

The pressure built until it found a weakness in the stone. There was a flaw in this rock, an inclusion of different material at its center, from which a fracture plane extended out across the matrix. Steam penetrated that flat seam of weakness, forcing open a microscopic separation.

The pressure flowed into pores in the rock like water seeking a path downhill. It poured into the fault plane there and further expanded the void at the center. As the inside temperature rose, more of the bound water flashed to steam.

When the strength built into the rock by its compression and its cemented structure was exceeded by the internal pressure, the bonds along the fracture plane failed and the rock split. The sound was like the crack of a rifle, but it was completely inaudible in the crescendo of the fire. An exploded puff of dust and vapor was instantly swept up into the gales of smoke. Water molecules that had been buried together in the space of a few centimeters for 70 million years were released in a fraction of a second. They shot off in every direction—swept away into the firestorm, invisibly dispersed, irreversibly blended with the air of the present.

THE SMOKE THINS as the wave of flames rolls past. The rounded rock is still in place, but now it lies divided into two halves. A fissure cuts vertically through the sandstone, exposing the center to the sky. The split reveals the structural imperfection that was the rock's weakness. It is the imprint of a single winged scallop shell, centered on the split rock face in perfect detail. The raised, parallel radial ridges of its shell stand out, etched with concentric growth lines. A scallop pearl—now a sphere of crystal quartz–

rests against the fossilized organs of the creature on the opposite face of the fracture line.

The perfect outline of the shell presents an image of a creature that no longer lives on Earth. Yet its ridged surface looks just like the living animal itself would have looked lying on an ancient beach.

The rush of the receding firestorm fades into the sighs of waves sliding down the sand into a Mesozoic ocean. That water is alive with the sea monsters of the Cretaceous—the sinuous reptiles that occupied the predatory whale niche before the whales did; the largest of the bony fish ever to swim the seas, living on plankton, occupying the baleen whale niche. Diving birds dart between these leviathans, looking like modern loons or cormorants—until they come close enough to reveal their teeth.

The Cretaceous scallops already had to deal with the primeval fish that hunted shelled prey—sturgeon, rays, and shell-crushing sharks. Those creatures have all changed through the ages, and followed a branching course along the evolutionary tree that charts their journey to our time.

The portrait of the scallop in stone came down to the modern era by a different route. It died long ago and was frozen in place, and frozen in time. Its rocky grave was rafted on floating continents, uplifted by rising mountain ranges, and rolled through a series of rivers. Its fossil shell now marks a point in time at which its species went extinct, but beyond which a host of other creatures survived.

A magpie alights on the fractured rock, looking to scavenge smaller creatures that did not escape the wildfire. The bird is the latest in a series of animals that have swum, then crawled—finally walking and flying through this canyon—and through all of its predecessor canyons—to sustain the spark of life through one change after another, all the way down to the present time.

— 6 —

Anting

THE BLACK GRACKLE sat on her perch, continually grooming herself. She acted uncomfortable, unbalanced, itchy. She had been drawing oils from the preen gland at the base of her tail and spreading them over her feathers until they were positively glossy. Nonetheless, she was in constant motion, distracted, jumping at small sounds.

As the day grew humid, an image came to her—she remembered that in the past, when she felt this malaise, she would throw herself upon a pile of ants. The bird knew all about the ants in these woods. There were scores of different kinds. She had a memory of one smallish species. It was not overly aggressive, not stinging or biting, and very good at micro-foraging. These ones usually nested in slightly mounded hives, in the open.

She glanced down but saw no open spaces that might support ant mounds. The view behind similarly revealed only the rolling contours of the forest canopy. Finally the bird dived away and wheeled through the canyons of green. She would to look for a meadow or a tree-fall gap. A mound of the little creatures might be waiting there to give her an ant-bath.

THE ANTS LIVED at stifling densities in their underground warrens. The corridors they plied were barely wide enough for two of them to pass each other in the darkness. Contagious diseases were a threat in such close quarters. But the ants fended those off through a symbiosis they had evolved with beneficial cultures of bacteria. The ants carried these micro-colonies of microbes in dents on their heads. Those particular bacteria worked for the ants—in exchange for the provision of the niche that supported them.

These bacteria produced a brace of different molecules. Those products spread out and coated each ant in a thin antimicrobial film. That coating allowed the ants to amble through their dank, airless tunnels all day and still stay healthy. They never came down with fungal or bacterial infections, in spite of all the spores floating all around them all the time.

Suddenly dust rained down on the workers just inside the hive entrance. The sunlight disappeared. Ants ran in every direction, antennae and mandibles raised, looking for a threat. They released an alarm pheromone. This signal diffused down into the hive, calling up more ants to join the confusion.

They swarmed out of the opening and found themselves lost in a forest of feathers. The strange blockade was thick enough to thwart their defensive fire drill. They slowed their running—the perceived assault appeared to have ended. None of them was emitting any more alarm signals. They all settled down to investigate the new blockage that had fallen upon their nest. It seemed that a recently dead animal had landed on their doorstep—a windfall for the hive.

They explored this new advent, walking upside-down on the imposed ceiling. The living surface was scattered with flakes of dried, shed bird-skin—almost pure protein. This was a discovery of far greater value than anything else the ants encountered on the trails they regularly followed. Further, they occasionally found louse eggs, a very high-value food source indeed. They even found some lice. Scouts that made these discoveries took samples of the unexpected bonanza back into the hive to recruit more foragers.

The ants explored between the tight layers of feathers, which were actually less oppressive than the hive conditions they were accustomed to. As they pressed between them, the flexible quills served as scrubbers that swept off and absorbed the oils from the ants' hard cuticle. After a while, the shiny ants had become dull and dry. Their waxy antimicrobial coatings had been transferred from their cuticle to the feathers and skin of the bird.

THE GRACKLE LAY sun-struck in the middle of the meadow. She was staring at the sky, halfway between ecstasy and panic. Her wings were spread out wide across the soil, mantling an ant nest. Her bill was open—she was panting in the heat of the still air at ground level. Her feathers were splayed, providing access for the ants to swarm over her skin. One eye or the other closed whenever an ant walked across that side of her head. Her feet were folded—she was sitting on the soil.

Her skin tingled with the sensation of the exfoliation process cleansing her all over. She squelched the impulse to flee from a swarm of bites and stings—even though these ants were not now aroused to attack. To a vulture lowering from above, she looked dead—feathers askew, covered in ants.

Finally the rush of sensation became too much to bear. She imagined a hawk had dipped behind the branches overhead while her eye on that side was closed. She hopped to her feet, glanced around the trees above, and burst into the air. Her flight path rose to match the curve of the wall of foliage at the edge of the clearing. At treetop level, she found a perch overlooking the meadow.

From this vantage point the bird scanned the area, but she saw nothing of concern. The hawk she had imagined turned out to be a vulture in the distance gliding against the high sky to the south. She realized that she was not concentrating on her surroundings—she was still distracted by ants crawling all over her skin. They showered from her wings when she shook herself. She smoothed her feathers, then scratched at an itch with one foot. But the sensations continued. Stray ants would be dropping from her body all day.

THE GRACKLE HAD left a trail of fallen ants scattered across the meadow. Many of them had landed within a few yards of the nest, shaken off when she jumped into the air. Others were still floating—standing weightless on the breeze until they ticked down on leaves in the trees, or in a trackless wasteland obscured in every direction by the bases of tall grass stalks. The farther away they fell from the hive, the longer it would take them to get back. Some of them would still be straggling home days later.

A few had fallen into the shade and came to rest on a stream at the meadow's edge. The flow carried them off until they chanced to find the bank, or they were seized by fish that leapt through the surface.

Fallen ants set off along search-pattern courses. They wandered across the scent trails of other ants, some of which they could not detect, others that they avoided. Eventually some of them came across the scent of a trail from their own species, and they bent their course to follow that. If the scent eventually diminished, they reversed course, to go in the direction in which the scent was increasing. In some cases, the trail led to another hive of ants from their own species. Most of the lost ants had lost their own scent when they were polished clean by bird feathers, but they were accepted into these foreign hives anyway.

The farther away from a nest they fell, the slighter were their chances of returning. The area was full of spiders, ant-lions waiting in conical fall traps, prowling lizards and thrushes. Still, the gains for the hive from the visit of the bird more than balanced the loss of a number of the hive workers.

OVER THE DAYS, the bird felt better. The microbial imbalance that had infested her skin no longer troubled her. The superficial rash had succumbed to the antibiotic oils she had picked up from the ants. Her fever subsided; her iridescence returned—her feathers shone like snakeskin.

As they died, some of the pathogenic bacteria on her skin burst open to release viruses. These inert little polyhedral crystals would rest in place, dormant. They would only come active to infect more of the same pathogenic skin bacteria, should those reappear and attempt to re-establish a presence among the bird's feathers.

On the other hand, some of the specialized bacteria that lived with the ants had been wiped off them and into the bird's feathers. Some of those would find their way to the bird's preen gland, and take up residence there. While they survived, they would produce the same antimicrobial oils for the bird that they had for their previous ant hosts.

The grackle grew more active and vigorous after her bath among the ants. She slept better and woke earlier after visiting their nest. As the days of spring brightened and grew longer, thoughts of a nest of her own began to stir in her consciousness.

NOTES

Many birds pursue a self-anointing anting behavior (Skutch, 1948). The presence of parasitic mites or microbial pathogens may drive this behavior. Beneficial mites (Doña et al., 2018) and microbes, however, are normal inhabitants of feathers. The avian uropygial gland at the base of the tail provides preen oils. In the hoopoe the gland supports symbiotic bacteria that add to those oils (Martin-Vivaldi et al., 2009). The preen-gland bacterial populations of the hoopoe vary over the seasons as needed, demonstrating some control over them by their avian hosts. Avian anting behavior may be driven by the recovery of ant antibiotics by the birds, for use on their own feathers (Ehrlich et al., 1986). Ants carry symbiotic bacteria that make antibiotic oils in the metapleural gland on their head (Fernandez-Martin et al., 2006). Birds will cover their feathers with oils, as has been illustrated in an example of grackles that groom themselves with mothballs (Clark et al., 1990).

REFERENCES

Clark, C. C., L. Clark, & L. Clark. 1990. Anting behavior by common grackles and European starlings. *Wilson Bulletin* 102, 167–69.
Doña, J., et al. 2018. Feather mites play a role in cleaning host feathers. *Molecular Ecology* 28, 203–18.
Ehrlich, P. R., D. S. Dobkin, & D. Wheye. 1986. The adaptive significance of anting. *Auk* 103, 835.
Fernandez-Martin, H., et al., 2006. Active use of the metapleural glands by ants in controlling fungal infection. *Proceedings of the Royal Society B* 273, 1689–95.
Martin-Vivaldi, M., et al. 2009. Antimicrobial chemicals in hoopoe preen secretions are produced by symbiotic bacteria. *Proceedings of the Royal Society B* 277: 123–30.
Skutch, A., 1948. Anting by some Costa Rican birds. *Wilson Bulletin* 60, 115–16.

— 7 —

Hanging Gardens

A BUTTERFLY PATTERNED with black veins on transparent wings drifts weightless just above the foliage. It ticks against the leaves, tasting them with its feet. Bold red chevrons centered on its hindwings flash in a patch of sun. The animal is looking for the plants that will host its next generation. Finally the insect settles motionless in a bunch of epiphytes. It rests on a leaf a hundred feet above the jungle floor. After a moment the butterfly curls her abdomen to place a single egg beneath the margin of the leaf.

She moves to another site a few feet away and touches down again. But there she draws the attention of an aggressive brown ant. The soldier lifts its head, antennae wide apart, then jumps at the visitor. The butterfly springs into the air and floats off on broad wings.

Later in the afternoon a smaller, more retiring ant wanders out of sight under a leaf. It discovers an egg stuck there upside down, grips it in its mandibles, and works it back and forth. Finally the egg comes free. The ant drops it away into space, and moves on.

These two ants, the larger soldier and the smaller worker, are gardening ants. Though they are members of two different species, their cultures have married together. They cultivate the same plant and cooperate in its protection. They both nest in its root-ball, high up in the branches.

Each of the two species contributes its own special skills toward the communal success of their joint effort. The smaller ones guard the leaves against insect herbivores. The larger ones defend against larger intruders. They also collect seeds to plant in the garden that the pair of species maintains. If these ants succeed in their efforts, the plants there will bear their purple fruit, and the ants will tear into it and harvest the seeds inside.

Ants from these paired species patrol the foliage day and night, but their challenge is greater after dark. Raiders come out of the blackness. They are announced only by the slightest puff of breeze, or the brush of weightless whiskers. Then one of the guards sprays a pheromone alarm into the air. The ants' night patrol comes alive. All the other ants raise gaping mandibles and race blindly in every direction.

However, their search patterns momentarily take them off the fruits and down the stem. Then they feel the bark they are running on recoil as the weight is removed from the end of the branch. When they run back up again they find only empty, moist pedestals remaining—bats have stolen the fruit they were guarding, along with all the seeds inside. That means the ants will have to go find those seeds and bring them back.

THESE GARDENING ANTS inhabit the seasonally flooded forest. They live among the roots of epiphytic plants high in the trees, above the high-water line. Their association forms the most successful social society in this lowland tropical forest habitat.

Both of the communal ant species collect bits of bark and dead leaf, as well as animal droppings. They bring this matter up into the trees and fold it in between the roots to fertilize the plants in their living nest. They create the earth into which those plants send the roots that hold them all up in the treetops.

The two ant species maintain separate brood chambers in the soil of their shared garden. The larger ants may occasionally bully the smaller ones— chasing them off the best food sources—but the two never actually fight. Their air plants depend on their cooperation to secure the sunny space in which leaves spread and flowers bloom. The ants, in turn, depend on the plants for shelter.

The larger ants prize the seeds of the epiphytic plants that grow in this patch. The living seed attracts the ants through its unique aroma. Even though they are hard and dry, these seeds maintain a constant fragrance. The larger ants descend from their nest to spend their days scouring the branches and the forest floor, searching for that aroma—the most enticing scent they know.

A handful of those seeds would smell faintly of sweet vanilla. They are dispersed far and wide across the forest in the droppings of fruit bats. But even after they have passed through the digestive system of a bat, their signature fragrance is still produced, enabling the ants to find them wherever they chanced to land.

The ants carry the seeds back one at a time, up the tall trunks, out along the branches to their garden, and then down through their tunnels into the dark. They place their harvest in the brood chambers close by their next generations—the helpless, white immature stages in the nursery. Perhaps the seed scent reminds the ants of their young, and they are following the same behavior they would in replacing a mislaid pupa or larva.

There the seeds germinate. The seedlings send their white shoots straight up to break into the light; and they send out their roots in every direction through the soil. Their fibrous sinews anchor the garden to the branches, and knit the fabric of the root-ball firm. And they also keep the soil there from growing too heavy with rain.

The suspended sphere of earth is drenched in the afternoon by thundershowers. The root-ball absorbs water and swells beyond twice its morning weight, bending its branch downward. The sodden soil between the roots threatens to liquefy and return to its natural level—to the forest floor. But the roots soak up water like a sponge. They draw it up into the main stem and distribute it out to the scores of leaves above.

The green surfaces are solar evaporators. The plant holds them up in the air, where the sunshine and the breeze can draw the water off into the sky. The net effect is to take water that weighs down the aerial garden and spread it out to dry over the thin, flat layer of several square meters of leaves.

It is their plants that allow these ants to live in an earthen nest high above the tropical wet forest. Moisture is so efficiently wicked away by the network of roots that it never penetrates to the hive center. The ants inside remain dry, even in torrential downpours that wash away workers caught on the branches outside.

STILL, THE GARDENING ants are not rooted forever to a particular aerial patch of soil. They are ready to move their nest when times change. Their optimal living space shifts through the contours of the forest over the years. They inhabit the most productive parts of the realm, such as the walls of foliage exposed to the light along the open corridors beside riverways.

Tree-fall gaps produce the most new growth, beneath skylights torn in the canopy where the oldest trees have fallen. A guild of early maturing vines and understory trees is adapted to seize the advantage of these temporarily bright spaces. Stationary sap-sucking insects settle in the profuse new growth, and the gardening ants in turn are attracted to those sap-tappers. The ants move into the area to farm honeydew, which is produced by scale insects and aphids that colonize the fastest-growing shoot tips. When they come, those ants bring their garden with them.

The ants notice the diminishing sunlight at their established nests when the sessile sap-sucking insects move away. When the light gap closes over and total shade returns, the local plant growth slows, and the ant colony splits. Half of the members march away, guided from tree to tree by scouts. They trail across tendrils and vines, up one trunk and down another. Their destination is a sunny glade that supports more sucking insects for them to shepherd, where the light is brighter and the new growth more vigorous.

Ants arriving at new sites find temporary bivouac among the creepers and moss, and then they set about planting the seeds of their favored epiphytic plants. Their hanging gardens will grow over the years, along with the other plants that eventually come to fill the gap in the canopy. When the gap closes over, the migratory imperative will cycle again.

The sedentary sucking insects grow as quickly as the shoots they colonize. The sap they tap is high in nitrogenous nutrients, which the scale insects and the aphids need to maintain their rapid reproduction. They absorb all the nitrogen they can get from the sweet sap; in the process, they imbibe more sugar than they need. Their stationary life does not demand much carbohydrate to provide them energy. They excrete the excess in droplets of honeydew.

It is sweeter than nectar, and the ants are quick to harvest it. Again, they divide their labors. The smaller ants patronize young aphid colonies, where the insects and their offerings of honeydew are smaller. When the sessile colony expands, the larger ants move in.

Their high-energy diet enables the gardening ants to pursue a high-energy lifestyle. They are alert and aggressive—they defend their trails and their sources of honeydew against other ants, which are scarce where the gardening ants are most active.

They actively defend their gardens. A monkey inspecting the hanging root-ball will find her skin crawling; the ants rip exposed skin with their mandibles and spray formic acid over the wound. They lift up the scales of reptiles that tarry too long in their domain, and jab their stingers underneath. Frugivorous grazers usually choose to pass up the fruits that adorn these gardens hanging in the sunshine. Leaf-eating lizards likewise soon leave.

NOW ANOTHER SMALL dusky butterfly drifts through shafts of sun above the foliage, looking to lay her own eggs. She is drawn to the tip of a shoot colonized by scale insects and guarded by patrols of gardening ants. The butterfly comes to a hover just above the colony; she bends her antennae down and probes the backs of the two closest ants. The sentries sense the faint sensation above them—the puff of a sudden breeze at their back, the brush of weightless feelers.

They respond by emitting an alarm pheromone; all the other ants lift gaping mandibles into the air and race in every direction. But when their search patterns take them away from one part of the colony for a moment, the butterfly descends, and in only seconds she places an egg on the stem, among the scale insects. Then she lifts off again just before the ants pass back.

The ants won't notice this alien presence in their flock—it is cloaked in pheromone-mimetic camouflage. And because of this subterfuge, they will also accept the tiny larva that emerges. And the larva will grow into a caterpillar, which will exude sweet offertory droplets just as the scale insects do. But this larval butterfly is no herbivore. It is a predator that prospers by eating the insects the ants shepherd. It is a parasite on the efforts of the ants, surviving and growing thanks to the work they do to protect the sedentary residents of this shoot.

The success of the gardening ants has created other resources that are exploited by a menagerie of parasites and predators. Some of these interlopers use their chemical mimicry to avoid notice while living right in the hanging gardens themselves. Others prey on the ants directly.

The ants are unable to defend their nests from swarms of army ants that file up the branches in endless columns. The invaders specialize in attacking arboreal social insects. The army kills the gardening ants in their tunnels in the suspended root-ball; then the invaders carry off the gardening ants' broods. The remaining garden-ant workers stand aside in the foliage. They will repopulate their homes when the invaders have moved on.

Other plants may take advantage of the hanging gardens as well. Some anthuriums and bromeliads that get a foothold in the garden may grow to out-compete the principal epiphytes. They can fill the root-ball with thick alien roots, and displace the ants from their own space.

But when the ants' gardens prosper, the entire epiphytic overstory benefits from their work. The epiphytic plants avail themselves of the soil which has been hauled up into the trees bit by bit and stabilized by the ant gardeners. Airborne fields of their flowers bloom in profusion high above the less productive, deeply shaded forest floor. Ant gardens abandoned to the shade are eventually populated by other arboreal arthropods, and overgrown by shade-tolerant orchids, lianas, figs and ferns—the emblematic decor of the aerial scaffolds in the tropics.

Nonetheless, among all the airborne growth, the particular plants cultured by the gardening ants make up more than half of the total epiphytic biomass in areas where the ants are active. The gardeners continually find places to cache the few seeds that will grow new gardens for them, to replace the old.

The two species of ants, and the select few species of hanging plants they cultivate, should all be able to survive and prosper independently—each on its own. Advantageous trade-offs that would commit them to a symbiotic life style are not obvious. But the gardening ants and their chosen plants are never found growing apart and alone. The success of their joint effort is reflected in their populations. Where their gardens flourish, the aggressive gardeners are the most prevalent animals in the forest—as anyone who chances to disturb their plants or their trails will soon discover.

IN NORTH AMERICAN temperate forests, we see a scaled-down version of gardening-ant mutualism in the *Trillium*. This plant spikes out of the shaded ground and unfurls a platter of three flat leaves a foot above the soil. Trilia produce a white bud at the central apex of their leaves. The bud may shade toward pink with age.

These trefoil plants are dispersed through their range by ants that live in the shade at the bottoms of north-facing slopes. The ants climb trillium stems in search of the hard black seeds, which are crowned with a cap made of the ants' favorite food. The ants take the seeds back to their queens, eat away the cap, and dispose of the seed in the bottom of the nest.

Trillium seeds lay dormant for many years. They are "double dormant": they pass two winters before germinating. Then they send down one slender root, but send up no leaf, the year they first germinate. Finally they send up a fine, slender primary leaf that drives the enlargement of their underground roots. Each fall the aboveground parts send their nutrients back underground and die off. Each spring, they send up a larger shoot, until finally they send up the three-fold symmetrical leaves that will host the central flower. A decade or more of growth may have passed before that.

From the flower blooming above an ant nest, the trillium will produce about fifteen seeds. Seedling survival rates will be small—deer are fond of trillium. But the few that do survive stand as reminders of the more extensive ant–plant mutualisms that flourish over the horizon, farther south, in the neotropical wet forests.

NOTES

Gardening ants are found in the New World tropics (Vantaux et al., 2007). The large garden ant *Camponothus femoratus* and the small garden ant *Crematogaster levoir* maintain separate brood chambers in their communal root-balls above the tropical lowlands. They share foraging trails, like the large and small leaf-cutter ants marching together through the upland forest. The gardeners feed on honeydew from sap-sucking insects, and on the extra-floral nectaries and pearl bodies produced by gap-filling trees (Davidson, 1988). Leaves growing from the gardens suck up moisture from the root ball and evaporate it off into the breeze at a rate that may average ten micrograms of water per square centimeter of leaf surface area per second. For a plant supporting a hundred leaves, each of which has a surface area of a hundred square centimeters, this amounts to the removal of ten ounces of water per hour from the earth around its roots, preventing the accumulation of sodden weight in its root ball (Yu, 1994). As it transpires, the plant absorbs carbon dioxide into its leaves from the air and moves dissolved nutrients up into its leaves from the base, driving its growth. The earthen tangle of roots maintains a consistency like cardboard, and the ants in the brood chambers within stay dry. *Peperomia* is a common gardening ant–associated epiphytic plant, along with certain anthuriums; other gardening ants live in bromeliad root balls (Cereghino et al., 2010); the garden usually contains several species of plants.

Ant garden plant seed is attractive to gardening ants, perhaps mimicking the scent of the brood. Ant garden–associated epiphytes make up more than half of the aerial plants in ant garden habitat (Nieder et al., 2000). These plants are rarely found growing in the absence of the ants that tend them (but see Morales & Vesconcelos, 2009).

One butterfly in the Metalmark family lays its eggs only in colonies of sessile insects guarded by the gardening ants (DeVries & Penz, 2000). The Metalmarks as a family are often associated with ants.

Trillium plants are found in the circumarctic forests of North America and Asia. Edible elaiosomes, which attract ants, are attached to their seeds.

REFERENCES

Cereghino, R., et al. 2010. Ants mediate the structure of phytotelm communities in an ant-garden bromiliad. *Ecology* 91, 1549–56.

Davidson, D. W. 1988. Ecological study of neotropical ant gardens. *Ecology* 69, 1138–52.

DeVries, P. J., & C. M. Penz. 2000. Entomophagy, behavior, and elongated thoracic legs in the myrmecophilous neotropical butterfly *Alesa amesis* (*Riodinidae*). *Biotropica* 32, 712–21.

Morales, S. C., & H. L. Vesconcelos. 2009. Long-term persistence of a neotropical ant-plant population in the absence of the obligate plant-ants. *Ecology* 90, 2375–83.

Neider, J., et al. 2000. Spatial distribution of vascular epiphytes (including hemi-epiphytes) in a lowland Amazonian rain forest (Surumoni crane plot) of southern Venezuela. *Biotropica* 32, 385–96.

Vantaux, A., et al. 2007. Parasitism versus mutualism in the ant-garden parabiosis between *Camponotes fermoratus* and *Crematogasater levoir. Insectes Sociaux* 54: 95–99.

Yu, D. W. 1994. The structural role of epiphytes in ant gardens. *Biotropica* 26: 222–26.

— 8 —

A New Color

THE GREEN JAY knew this quiet hollow well. She was an inquisitive, observant denizen of these woods. She had spent the early season here foraging through the leaves, but much of that time was actually spent just watching. Perched concealed in the branches, she noticed the animals that came through. She was looking for the predators—the snakes, coatis, and hawks. She roosted here at night. After weeks in residence, she finally began to search for nesting materials.

There were several sturdy nesting spots suspended in the forks of the branches. She avoided them all. She was a mature bird, with years of experience in picking sites to raise a brood. She had learned by trial and error which sorts of places would succeed.

She built her nest behind the walls of foliage, in the deep shade, beyond the sight lines drawn by the light and shadow. When other creatures were present, she stopped her construction and moved off. She built a second, poorly concealed sham nest higher up. She watched it for visits by aggressive toucans or monkeys. Her mate was perched in the scaffolds far above, out of sight. His calls to other jays filtered down to her. They fill the clearing with a chatter that would draw opportunistic bird hunters toward the higher levels of the canopy.

AS THE SEASON advanced, the female no longer foraged through her nesting hollow. Now she hunted some distance away, out of sight. She carried eggs within her—they were just about ready to be added to the nest. But as she looked for bugs and spiders one afternoon she heard a distant rush of sound off in the direction of her clearing. Like a tree-fall, it was a sudden gust on an otherwise calm day in the forest. The last of the outburst was punctuated with the cracks of breaking timber, then the quiet returned. She listened for a while, then flew back to investigate. What she found was chaos.

The sleepy hollow had been transformed. The floor space was now filled with a large, spreading bough dropped from above. Bits of dust and lichen settled through the air. Tilted leaves had come alive. A menagerie of all sorts of creatures—usually shy and well camouflaged—were looking to return to shelter. Katydids, walking-sticks, and lantern bugs clambered unsteadily across fallen foliage. A wary brown iguana limped back up the side of a tree trunk.

The large falling scaffold had sheared away a lower branch and exposed her nest site to the sky. The nest she had built was gone—eventually she noticed it face-down on the forest floor below. She did not linger long at the scene but continued moving. Animals drawn to hunting opportunities presented by tree-fall sites would be arriving soon. She had lost many a nest before, usually to the depredations of nest robbers. She took wing and fled.

SEVERAL DAYS LATER the green jay returned. The eggs within her had paused in their development—they would wait for a new nest to be completed. She had hastily found an alternative nesting site, and her mate had altered his territorial center accordingly. His call from high in the trees was now muted through distance.

She was back looking for building material. Much of it would be most easily gathered from her own previous nest, now downed in the clearing. She untangled sticks from the abandoned weave and flew away with them, returning again a few minutes later for more.

As she glided down across the open space, the jay noticed something—a glimmer of light—close to the ground. It appeared suspended in the roots at the base of a giant ceiba tree on one side of the hollow. Her innate curiosity about all things in her habitat bent her course, and she landed in the branches above the luminous apparition.

The object stood out from everything else. It was a different color—a color only birds can see. But it was out of place. It was a color never seen in the forest. Usually it was seen in the sky.

A string of clicks escaped from her throat. It was a voice that comes spontaneously to jays, when something unusual shows up in their environment. It was not loud, but the air was calm, so her sharp-eared mate could hear it nonetheless, from a hundred yards up and away.

She chose a lower branch, and dropped down for a closer look. As she perched on the lower site, she glanced over at the source of light, and immediately leapt back into the air with a reflexive startle call. The color had brightened at her approach, as if warning her away. She flew to a higher branch and looked back to find that the source was dim again.

Her mate heard her alarm and followed the sound of her mutterings through the wood. He landed in the branches above and called to her, but she did not fly up, so he came further down. She was looking at something. As he descended, he saw it too. It shimmered with that faint color only birds can see, but right there near the ground. A trill of clicks emerged from his throat.

The two birds began calling to each other—an expression of mixed excitement and anxiety—as though they were mobbing an owl. These birds prospered through intimate experience with all there was in their forest. They taught themselves what could be eaten, what to challenge, and what to avoid. But here was something they did not know.

They watched each other approach the unknown object, then jump back. They measured each other's confidence and trepidation by the inflections in their voices. High above them, other jays were descending from the canopy, drawn to the commotion from far and wide, their calls rising with a questioning note. Soon the clearing was filled with jays.

The birds were all driven by the tension between curiosity and apprehension. What was this new thing? Their only answer was to bounce back and forth from limb to limb and chatter to each other. They approached the object, then flew back. In the excitement, pecking-order confrontations arose when the birds intruded on each other's space.

Eventually the senior birds forgot their fear. They recognized that the light source was inanimate. They grew accustomed to its welcoming increase in brightness as they approached. None of them perched upon it, because when they drew close enough, its outline blurred in the luminous halo that surrounded it. But a few stood before it; the boldest among them gave it a peck. The impact sounded with a clink. The material was harder than wood, harder than sandstone, harder than anything they knew of in the forest. They had never before experienced a metallic object.

THE STORY OF this rock began long before there were birds in the forest—before there even were birds, or any other animals. Before life came out of the ocean the rock was already formed—born before there was any life on Earth at all—in the cataclysmic demise of a young planetesimal beyond the orbit of Jupiter

Back then, the orbits of objects circling the sun were not circular. Planets and smaller planetesimals flew through the outer solar system in wide ellipses. These spheres had come into being containing molten metallic cores. Convective currents within those cores generated magnetic fields similar to that of Earth. As the cores cooled, they formed solid iron. When they solidified, convection ceased and their magnetic fields weakened. All that remained was the magnetism that had been induced in the solid iron when the planetary magnetic field was still alive.

When they passed each other, the gravities of the largest of these planetesimals threw the smaller ones into widely elongated orbits. Those objects flew around each other in ever-changing trajectories. Eventually pairs of them came together in silent head-on collisions that pulverized both bodies into dust and rubble. Included in that debris were shards of their shattered iron cores.

The fragments that arose from these collisions went spinning away in random directions, each sailing through the solar system in its own new orbit. Eventually most of the debris would fall into the sun or be swept up by the surviving planets as they grew ever heavier.

Fewer and fewer of those tumbling shards—including those made of magnetic iron—remain free-floating. But occasionally, some of them still come down in the inner solar system—down to Earth, even today.

BIRDS CAN SEE magnetism. Molecules that respond to magnetic fields rest on their retinas. We don't know just what the magnetism looks like when it appears in their visual field. It is a sight we can't experience, because we cannot see magnetism.

The core of the Earth is still molten. It still generates a magnetic field, which the birds can see. Its lines of force circle the globe. The birds perceive those as latitude lines in the sky aligned with magnetic north. These visible bands serve the birds as guidance cues for orientation and migration.

Avian magnetoperception may rise and fall with the seasons. The lines in the sky may fade during the months between migrations, then strengthen again with the approach of spring or fall. Nonmigratory birds may not see them at all.

But most birds have migratory aspirations—some farther than others. To be useful, their magnetic sense has to be acute, because the Earth's lines of force are relatively weak. The field close to a magnetized piece of iron is stronger than the background strength of Earth's magnetism. Migratory birds guided by Earth's magnetic field can be disoriented en route when they fly past magnetized monoliths or magnetite deposits in the ground.

The meteorite of magnetized iron that fell from space and through the trees landed in a spot where the birds could approach it close enough to touch. It fairly gleamed in their magnetic sight. The closer they came to it, the brighter it glowed.

Birds without much curiosity—the smaller passerine birds, nonmigratory species—would pass by the metallic rock without pausing. Many of the other passing creatures—butterflies, day-flying moths, dragonflies, salamanders—could see the special color as well, but they too did not pause to take notice. The lizards, the rodents, the anteaters did not alter their paths at all. They were blind to its color and paid no attention to the rock resting in the buttress roots of the ceiba tree.

But others of the dominant birds, who were naturally curious about their environment, gathered to fill the clearing with shouts to each other and flights back and forth. Oropendolas, grackles, or toucans would show up in noisy conventions to fill the jungle halls with their calls for five or ten minutes. Then they would diffuse away, their voices fading into the trees, and the scene would return to quiet.

The birds could see the rock glowing at dusk, before they retired for the evening. If leaves fell during a windy afternoon and covered the rock, the birds could still see the spectral glow. One of them would descend to the spot, scratch away leaves with its talons, and pull off the last of them

with its beak. Then for an instant it would be face-to-face with the rock glowing at its brightest before the bird would call out and leap away for the branches.

Over time, the birds became accustomed to the special rock. They accepted it as a feature of the clearing, and they came less often to gawk at it. Eventually it would sink out of sight through the leaf mold. Its magnetic field remained intact, even as its surface rusted underground. But as it receded through the distance of its burial, its glow in the sight of the birds became more and more tenuous.

The tree-fall gap in the forest where the rock had crashed to Earth eventually began to fill with cecropia trees. The edge of the clearing retained its faint tint of the color that birds see in the sky. But the hollow lost its interest as a gathering place. The clearing shined on rainy evenings in the cold fire of luminous fungi growing on the fallen branches. Phosphorescent fireflies and click beetles still drifted through the area at night. But the base of the ceiba tree no longer commanded the birds' attention with that special magnetic glow.

NOTES

Birds can perceive Earth's magnetic field (Rodgers & Hore, 2009) and use that perception to guide their migrations (Engles et al., 2014). There are molecules of cryptochrome on their retinas. These molecules have the capacity to respond to weak magnetic fields (Fusani et al., 2013). (Humans also carry cryptochrome on their retinas, as do some other terrestrial organisms). The location of cryptochrome on the retina suggests that the birds perceive magnetism as a color in their visual field. The birds see colors beyond our visual range; we cannot imagine what they look like, since we cannot form images in shades of color beyond the ones we know. We will likely never know what color they perceive when they sense magnetism.

Planetary magnetic fields arise from the electric fields generated by dynamic convective flow in the planets' molten cores. The core of the Earth formed the same way as the molten cores of the planetesimals did long ago. Earth's iron inner-core has now solidified, whereas Earth's outer core is still molten (Bono et al., 2019), still radiating away its heat to space through volcanic vents on the surface, and still generating a magnetic field around us. The strength of a magnetic field diminishes with the cube of the distance away from the magnet. Measuring that strength optically, the brightness of a magnet would appear to jump in brilliance eight fold when a bird closed its distance to it by half. Earth's surface magnetic field is on the order of 0.5 gauss; magnetic iron can have a local field strength of hundreds of gauss; magnetic iron in meteorites called pallasites has been measured at 1.5 gauss (Bryson et al., 2015). A pallasite once fell near Pallasovka, Russia. Both that class of meteorites and that town are named after Peter Pallas, the German naturalist who first studied magnetic Russian iron meteorites. (The asteroid 2 Pallas, however, is not named for Peter Pallas; its name derives from Greek mythology.)

REFERENCES

Bono, R. K., et al. 2019. Young inner core inferred from Ediacarian ultra-low geomagnetic field intensity. *Nature Geoscience* 12, 143–47.

Bryson, J. F., et al. 2015. Long-lived magnetism from solidification-driven convection on the pallasite parent body. *Nature* 517, 472–75.

Engles, S., et al. 2014. Anthropogenic electromagnetic noise disrupts magnetic compass orientation in a migratory songbird. *Nature* 509, 353–56.

Fusani, L., et al. 2013. Cryptochrome expression in the eye of migratory birds depends on their migratory status. *Journal of Experimental Biology* 217, 918–23.

Rodgers, C. T., & P. J. Hore. 2009. Chemical magnitoperception in birds. *Proceedings of the National Academy of Sciences* 106, 353–60.

— 9 —

Sunfish

THE MORNING WAS placid enough, but the juveniles in the raft of gulls were still growing agitated. Stippled in their grays and browns, they were calling back and forth. In contrast, the less excitable adults, in their mature black and white, remained quiet but watchful.

They were all sitting in the sun, rolling with the swells on the open ocean. But with their heads held high, they could not easily see what was going on below. The juveniles felt like sitting ducks, vulnerable to attack from beneath by a breaching marlin or shark. And now the water was brightening all around—something had seen them there on the surface and was rising up from underneath.

Soon the yearling birds leapt squawking into the sky and headed for the horizon. The adults were not as flustered. They had not seen the torpedo shapes of the predators that threaten floating birds. Something was definitely coming up toward them, but it was coming gradually—like the flat top of a round column extending vertically from the depths.

The experienced birds held their positions. They were calm, but they were also wary of anything touching their feet, hanging below where the birds could not watch them. As the water continued to brighten, they all finally kicked into the air. One by one, they banked into wide, circling turns. They cried out to each other, looking down from above on what was coming up from below.

The top of the rising pillar was a circle at least ten feet in diameter. Pointed, flat wings extended from its opposite edges. From the gulls' new angle, they could see it was unsupported—the dark, steep sides were just its shadow.

The water sighed and ran off as the disk broke the surface. The parting waves revealed a huge, broad fish. This animal usually swam vertically—a discus suspended from its upper edge, but now it was lying awash, flat on its side. The birds watched, keeping their distance above, calling back and forth, now and then. But the fish did not move. It just floated there as if it was dead—staring blankly up at the sun, inviting them to touch down.

So the inquisitive gulls descended. The fish was very stable—it weighed two tons, and was too big to be moved by the swells. The birds were reminded of the firm ground of an island, though the nearest shore was miles away. One of the gulls off to the side put down its feet to slow its flight. It flared to a stop, and settled through the air to touch down on the creature. Nothing happened. The bird folded its wings and looked around. This was the first time in days it had stood.

The living island was a sunfish. Riding on its broad expanse was like walking in a tide pool, but the surface was softer—not encrusted with coral or shells. The body was grooved and cratered where earlier birds or cleaning fish had dug holes in it. A large pectoral fin lay folded down flat across the highest point, toward the center of the wide plate

After the example of their leader, the other gulls touched down. Standing on the mucus-covered tail fin was like standing on the mud in a tidal flat. Walking around was like foraging across a beach. And the animal life to be found here was more accessible and more diverse than on most exposed reefs or sand bars.

Fish lice prospered in this habitat, standing out like overgrown pill bugs fixed in place. Small arthropods crawled around like roaches on fallen fruit; larger ones were rooted, their tops dangling like tassels. Black, blood-red, or tan spiny-headed worms and flatworms infested the thick skin. And, to the birds' excitement, all of these things were edible, and there for the taking. With the appearance of this sunfish, a promising day of foraging had presented itself out of the deep blue.

THE SUNFISH IS a movable feast. It lives on a diet of jellyfish, the only things the ponderous animal can catch. It has to eat tons of them—they are mostly water, and low in nutrients. The great size of the animal—the largest of the bony fish—is a testament to the flourishing population of its jellyfish prey in the open ocean.

A mature sunfish sails at one knot, with a turning radius of fifty feet—it is a creature of wide-open spaces. As its mottled skin glides past, hitchhiking parasites can easily snag a ride and then populate the living terrain with their next generations. The parasites roughen its surface or hang into the water, adding resistance to the big fish's efforts to swim. The fish has no way to rid itself of such infestations, except that sometimes it can call down the birds for relief from its burden of parasites.

Standing on their floating island, the gulls had a new perspective on their ocean—from several inches higher. They could see other high-flying birds descending, gliding down to investigate the light patch on the water on which gulls stood.

A mixed flock of plovers and curlews in the middle of a trans-Pacific migration dropped in, with much chattering to each other. They landed on long legs at their unexpected rest stop and soon began probing into the surface, feeding between their toes. Sea birds of other species flew past, banked around, and then they too touched down and commenced their own prospecting. Other web-footed foragers—grebes, a lone albatross—lit on the water to pick around the edges of the basking beast.

The juvenile gulls were returning. One young bird landed near the front of the fish, and found itself face to face with a translucent dark-blue hemisphere the size of its foot, embedded in the surface. The bird cocked its head for a closer inspection and could see its own reflection on the glistening disk. But then its sharp beak drew too close. The limpid ball twitched, bringing a darker pupil around to focus on the bird with its fixed stare. The gull, realizing this was an eye, jumped back into the sky, crying his warnings once more.

The older birds watched him fly away again, then they went back to their own pursuits. Many of them were already on one foot, looking at each other, crops full. All the while, the floating fish continued to rest on the surface at the center of the avian convention. One by one, the itchy spots on its flank finally knew relief. While the smaller creatures worked over its skin, the animal, true to its name, lay placidly sunning itself.

WHEN THE SUNFISH carries its parasites into the midst of a raft of gulls, it provides a node in a food chain that connects the deep sea with the sky and land. The gulls are the links between the three states of matter there. The birds are equally at home slicing through the gusts, bobbing on the swells, or walking on the seashore—or on a sunfish. They connect the energy of the oceanic food chain to all the other places in the world they visit.

That food chain loops around into a cycle. When the gulls eventually die, often on the shore, the essential elements of life harvested from the ocean will leave their bodies. Those elements will eventually find their way back once again to the sea, whence they came. And there, they will be absorbed by plankton, which will in turn be captured by jellyfish.

FINALLY THE WATER began to rise up around the ankles of the gulls on the lee side of the fish. The rising tide floated them off their feet, and they found themselves paddling once more on open water. Their broad platform slanted more steeply downward, shading to blue as it sank away. The surf rushed up and over the fish's eye, which darkened as it adjusted from bright sunlight to the dappled, subsurface vista. The fish was returning to its vertical posture.

The curlews and plovers called out in unison and sprung into the air to rejoin their migration. The gulls swam clockwise around the vortex left in the water by the sinking sunfish. They lingered in the area where the beast had disappeared into the depths, watching to see if, perhaps, the animal might reappear and present its other side.

NOTES

Many of the largest animals in the sea feed on the smallest animals in the sea: plankton. Plankton is defined as those drifting creatures that cannot resist the pull of the current. By this definition, jellyfish are the largest of the plankton. The sunfish, one of the largest of the fish, feeds on them.

There are more than fifty documented parasites on the *Mola mola,* the Ocean Sunfish. Its outer surface crawls with isopods, copepods, flatworms. Other crustaceans, such as the fleshy barnacles, or the copepod *Pennella,* are immobile—firmly attached. Stately, slow-moving adult sunfish are heavily parasitized (Love & Moser, 1983) and often seek out other creatures to clean the external parasites off their flanks. They also carry a heavy burden of internal parasites, which parasite-cleaning birds and fish cannot reach.

REFERENCES

Love, M. S., & M. Moser. 1983. A checklist of parasites of California, Oregon and Washington marine and estuarine fishes. NOAA Technical Report NMFS SSRF-77, 432–36.

— 10 —

Pillar of Life

A MANGROVE SEED head floats in place, just below the waterline. The seedling resists the pull of a rising tide—its primary root, reaching straight down, has snagged on a shallow silt reef. The anchored head leaves a V-shaped wake on the open surface.

The seas offshore have been calm for a week. The plant is establishing a muddy-bottom foothold. It is setting the stage for a new habitat to rise out of the ocean. The plant will produce a bulwark of roots. They will be stout enough to hold their ground against any weather that may ride in on the trade winds from the east.

This mangrove seed had germinated months before it found its muddy base. It had been one of many spikes suspended from the mother tree. The spikes were pendulous seedling roots, already germinated and hanging tip-down. Those seedlings had grown a foot long before they fell from the branches.

After the seedling dropped, it floated for a week or two, then its buoyancy changed. Its root-tip end sank. It rode on the current hanging vertically from the surface, trolling for the bottom. When it chanced upon shallow waters, the seedling's lower end dragged across the mud like an anchor. This slowed its progress on the current.

These seedlings move most slowly when in touch with the silt. So they accumulate over the crests of mud bars or tidal flats. They trace tracks back and forth across the soft bottom as they ride in and out on the tide. When they have run hard aground, the young plants lose still more buoyancy. They establish their footing in place.

The depth through which a floating seedling can touch the bottom determines the eventual boundaries of the mangrove. The forest establishes itself where the water is no deeper at high tide than the length of a seedling. Those seedlings that run aground farthest from shore will find their first leaves flooded with the highest tides. Some will be uprooted from the more seaward reaches of the shallows. But if stormy seas stay away long enough, rooting spots on the landward side will establish clumps of mangrove shoots. In the years that follow, other seedlings will flourish in the lee of the established plants.

Mangrove buttress roots becalm the water between them, so when the current slows under clear weather, particles carried on the tides will settle out in their shade. Together with sunken mangrove leaves, this rain of sediment will start to raise the bottom. The higher it rises, the more drifting seedlings from other mangroves will find anchor there. The increasing population of plants will expand to become a mangrove islet. Eventually, dry land will emerge at its center.

The mangrove islet will create living spaces much richer than the wave-scoured tidal flat that was there before. As the habitat stabilizes, prospective residents will arrive from all directions—from the land side and from the open ocean. Intertidal creatures will be joined by insects, snakes, arboreal lizards, and crocodiles. Terns and boobies will drop down from the air to nest alongside raucous long-legged herons. Below the waterline, shellfish will encrust the roots and finfish will convert the shady recesses into hatchery and nursery space.

Over the decades, a sunny, silty bottom once braided by the currents becomes a fertile, stable world, neither sea nor shore. There is no beach or rocky margin to demarcate the transitions between the mangroves and terra firma on one side, or the ocean on the other. The offshore forest merely grows denser, eventually expanding shoreward to enter into swamps and river deltas.

The mangrove swarms with life everywhere—the denser the tangles, the greater the diversity. Many of the creatures that live there are unique, having given up their previous identities and adapted to the niche in between the sea and the shore. The impenetrable growth teems with tenants of all stripes—even in a single shaft of sunlight falling into a still tidal pool through a break in the leafy ceiling.

THE DEEPEST HEART of the mangrove forest is a shadow world under a closed canopy. The tide ebbs and flows up and down the prop-root trunks in synchrony with the daily passage of the moon. The acreage is broad and silent—a jungle so thick that the sound of the ocean fails to penetrate.

But there are voids in the thicket of stems. Where one larger cavern opens out into the foliage, a spherical chamber is formed with the waterline as its equator. Its walls of branches rise and converge above the pool; its tangle of roots descends through the surface to fill the water below. A single opening in the overhead canopy remains over the center of the pool. A shaft of sunlight coming through that central opening illuminates the vertical axis of this chamber. It passes through the water surface and continues into the darkness farther down.

As the sun climbs above, this room in the mangrove brightens. The shaft of light coming down from top center is outlined in a dancing swarm of sparks. These floating motes are midges, hundreds of them, each only millimeters long. They are invisible when flying through the shadows—too small to see—but when they break the virtual edge of the pillar of sun, they come alight.

The tiny fairies can see the sun above when they move into its column of light. And they can see the light reflected from their sunlit peers flying ahead of them. Their intent is to be seen—this is a mating swarm—and when they are all in the limelight, the chase is on.

From the surrounding shadows they home in on each other's firefly forms. If they break out of the beam, they bend their flight back around, like moths drawn to the flame. They hover or zoom off in pursuit of one another, animating the narrow cylinder of light.

Birds that catch insects in midair visit this living cavern, but they are thwarted by the midges' schooling behavior. One mote of light replaces another, to be replaced by yet another in the sights of a gnatcatcher or a warbler. The pond-hawks cannot single out one target to snap from the air. The midges' swarm is an effective defense, and it is replicated where the light passes through the waterline, to light other glints of life farther below.

THE MIDGES SOMETIMES bottom out, dimpling the pond with tiny single ripples where they bounce off the water. The shaft of sunlight leaves them behind when it continues through the surface. It slants down into darker water where once again it brightens in the reflected glow of material suspended in its path. As above, that material is alive.

The orbit of midges above the surface is mirrored by a school of plankton orbiting in the pillar of sunlight below. This is another mating swarm, a congregation of copepods—thousands of them, each less than a millimeter in length. They are invisible in the brackish water—too small to detect. But when they swim into the narrow sunbeam below the opening in the mangrove dome, they brighten. Then they can be seen, and they can see each other. They home in on one another's glow and fill the column with their reflected light.

The copepods prey upon smaller plankton, and are in turn prey for planktivorous fish, but the highlighted school does not attract many of its own predators. Long silver snappers and juvenile barracuda wait like loaded crossbows, station-keeping in the shade of the mangrove roots. They are looking for any smaller fry that might be lured from cover to the column of fish-food teeming in the pool's center. Under the watch of their guardians, the thin vertical swarm of copepods goes unmolested.

BUT THE SWIMMERS in the stream of light are not free of predation. The beneficiaries of every survival strategy are eventually exploited through the rise of a counterstrategy. The strange organism that has arisen to fill the predator role here is a creature of the mangrove. It long ago gave up its sea-going ways to adapt to this in-between habitat. It stalks its prey from within their shining swarm. And though the column of plankton stands out in the shaft of sun, the predator in their midst is invisible.

A close look into the school of copepods shows that some of the creatures are motionless. They lie at angles, heeled far over from the posture of most of their swimming kin. But these hapless animals do not sink away from the crowd—they trail through the swarm, even though their oar-legs are frozen. They are puppets, pulled by invisible strings.

From some angles, a void appears to be moving through the sunlit cloud of plankton, ahead of these paralyzed individuals. With the right light, the edge of a sphere appears in that void—a prismatic arc of light refracting around its virtual circumference. But most of the time its outlines are obscured by the mass of living plankton it spends its days within.

The diaphanous predator is a box jellyfish. Its bell is a centimeter in diameter. It hunts by trawling its four fishing lines behind it through the shaft of light. This snare is unaffected by the confusion-defense of the plankton's schooling behavior. The jellyfish moves through the swarm guided only by the light, trailing prey snagged on transparent threads that stream from the corners of its cuboid bell.

UNLIKE JELLYFISH DRIFTING passively on the open water, the box jelly-fish are purposeful swimmers. They move against the current with heartbeat pulses of their bells. They do not trail long streamers that disappear into the depths; in the close confines of the shallows, they hunt for both freshwater and saltwater prey. They are aggressive patrollers of estuaries, river mouths, bayous. They navigate around obstacles, hunting through the shade of man-groves or under palm trees that reach out over the inlets along the shore.

These jellyfish carry sets of different eyes, some of which have lenses and retinas; they appear to distinguish shapes and individual points of light. Their adaptation to the shoreline environment has made them as advanced over the rest of their jellyfish kin as the octopi and squid are advanced over the rest of the mollusks.

Many different species of box jellies hunt along the subtropical seaside habitats of the world. The ones that hunt copepods in the mangroves are among the smallest. A dozen of them would fit in the palm of your hand, their stings only a subliminal tingle. Others, such as the Australian box jellyfish, grow to the size of a beach ball, with streamers trailing ten feet behind. Their sting can penetrate the hard shells of crayfish and is venom-ous enough to kill humans.

All the box jellies are weightless and delicate, protected only by the fiery shock of their touch. Some of them appear to recognize their kin and orga-nize themselves into hunting packs. They advance in a line, combing the shallows with a dragnet of lethal tentacles. They are searching for anything that swims or crawls through salty ocean or fresh estuarine water. Being transparent, they are difficult to avoid. The sting of a big one is as unan-ticipated as that of a mad hornet that drops unbidden from the air—thus their common name, sea wasps.

THE DAY EVOLVES and the slanted shaft of sunlight moves across the pool in the cavern in the mangroves, its axis pivoting to the east. The split col-umn of highlighted midges above the waterline and plankton below moves along in parallel, tilting ever more eastward into the afternoon. As the angle

becomes shallower this pillar of life grows more crimped, its axis bent at the waterline by refraction. Later in the afternoon the light breaks up into dapples and is finally cut off by the overarching canopy.

Shades of evening warm the leafy dome above as the daylight fades. Without the sun, the swarming creatures in the room in the thicket of mangroves lose their bearings and diffuse off into the shadows. The afternoon breeze dies away, and all activity winds down with the approach of the dusk.

Water levels drop around tree trunks as the high tide starts to ebb. After sunset, moonlight rises in the east to make silent silhouettes out of branches and buttress roots suspended above black water.

In the darkness, the unidentified sounds in the mangrove include occasional knocks along the waterline. Those are mangrove seedlings floating on the current. They collide with the prop roots of their parents as they percolate through the forest on the outgoing tide. Lying flat on the surface like floating daggers, they spin through eddies, slipping past more quickly the faster the water level drops.

The long seedlings eventually leave the suspended forest and float through the estuaries along the shore, or out to sea. They will live long enough to drift across the ocean and colonize continental margins and islands all around the world. The mangrove forests stabilize the shore behind them. And within their own shade, they support a network of creatures from land and sea that have adopted this netherworld as home.

NOTES

Mangrove trees construct and maintain their own habitat on the warm-water tidal-flat shoreline. Their adaptations to this life include breathing straws that rise from the mud to supply air to their flooded root system (black mangrove), or stilt-like prop roots that hold the air holes (lenticels) on their trunks above the waterline (red mangrove). Most plants drop seeds that are dormant, but mangrove trees bear green seedlings, with a descending taproot already emerged. A cadre of creatures lives among the mangrove branches, including the mangrove warbler, the mangrove flycatcher, and the mangrove vireo. Among the mangrove's immersed roots, the small mangrove box jellyfish (*Tripedalia cystophora*; Buskey, 2003) preys upon copepods.

The copepods (e.g., *Dioithora oculata*) are phototropic and can accumulate in light shafts under Red Mangroves (Buskey et al., 1996) in swarms of a hundred organisms per milliliter of water (Stewart, 1996); they organize their swarms using visual cues (Hamner & Carelton, 1979). Their swarms in the mangroves may survive due to the suppression of most of their planktivorous predators by the more dominant fish waiting among the prop roots (Buskey et al., 1996).

Their jellyfish nemesis is an aggressive predator that actively hunts and navigates by sight, as do the other members of its class (the Cubozoa; Garm et al., 2011). One type of eye they carry hangs pendulous near the bottom of the bell, on a weighted stalk. This pendulum orients the eye's view skyward no matter what the creature's posture. These spherical eyes take in a wide-angle vista of the sky. They orient the creature's pulsing resistance to the current by keeping the break in the leaves centered directly above, holding the box jellyfish in the light (Garm et al., 2011). The lens of that jellyfish eye gathers the light coming from a 100° cone directly above. But the water bends that light (bending it more the farther from the vertical it falls). The jellyfish uses the water like eye glasses; refractive bending (described by Snell's law) gives the pendulous eye a view spanning 180°—the entire sky above the waterline (Garm et al., 2011). The advanced nervous system of box jellies supports not only vision but also their aggressive swimming pursuit.

The largest members of the family (e.g., *Chironex flickeri*) have bells the size of a basketball and streamers ten feet long. They are delicate creatures, not found over coral or rocky bottoms. They would be torn apart by the prawns and crabs they attack, were their prey not immobilized so quickly. These jellies hunt the waters of Australian swimming beaches and spawn in river mouths. There have been sixty-three recorded fatalities attributed to *C. flickeri* around Australia (Williamson et al., 1996).

REFERENCES

Buskey, E. J. 2003. Behavioral adaptations of the Cubozoan medusa *Tripedalia cystophora* for feeding on copepod (*Dioithona oculata*) swarms. *Marine Biology* 142, 225–32.

Buskey, E. J., et al. 1996. The swarming behavior of the copepod *Dioithona oculata*: In situ and laboratory studies. *Limnology & Oceanography* 41, 513–21.

Garm, A., et al. 2011. Box jellyfish use terrestrial vision cues for navigation. *Current Biology* 21, 1–6.

Hamner, W. M., & J. H. Carleton. 1979. Copepod swarms: Attributes and role in coral reef ecosystems. *Limnology & Oceanology* 24, 1–14.

Stewart, S. E. 1996. Field behavior of *Tripedalia cystophora* (class Cubozoa). *Marine and Freshwater Behavioral Physiology* 27, 175–88.

Williamson, J. A., et al. 1996. *Venomous and Poisonous Marine Animals: A Medical and Biological Handbook*. University of New South Wales Press, Sydney.

— II —

Chickaree Delivery

THE PINE SQUIRREL rested in his tunnel on a lumpy bed of nuts—walnuts, chestnuts, a scattering of hazelnuts, and some hickory nuts—his stash for the winter. Earlier, he had ventured far down the hillside, right to the border of the territory of an older, more aggressive squirrel. There, he sniffed out fallen acorns to add to his cache. Now the scent of those acorns in this den filled his head. He would probably eat them first—when the coming cold set in. He had already stopped searching for whatever other forage the forest might still conceal.

He sat in the dark in his excavation under a log, which was itself buried under a pile of pine cones two feet deep. He had spent his summer afternoons high in the branches. He cut down those cones, then ran down the tree trunks to retrieve and carry them to this hoard.

He had piled them against a dead tree stump, between a pair of fallen logs. This kept them in the shade, where they would not dry out, open up, and release their pine nuts. He would be counting on those nuts to sustain him through the rest of the cold season ahead. All summer he had worked to amass this cache, and now he would defend it with his life against all comers.

The morning air outside was chill—fall would be a short season here. The squirrel's ears had already grown their winter tufts. In the branches far above, no pine nuts remained. The scales on the smaller cones he had left in place had dried and opened wide. All the nuts inside had floated off, spinning away to disperse across the forest.

There was not much left at ground level either. No more mushrooms or truffles waited to be found beneath the undergrowth. The squirrel had searched out all the fallen forage in his territory. This even included some nuts that had been resting in the grass for years and had lost most of their scent. He was now done searching—he would be spending the next half of this year right here, hidden in his midden.

THE SNAP OF a twig brought the squirrel's head up. He looked around to focus his ears toward the mouth of his tunnel. The sound of another, bigger branch breaking brought him to his feet. Immediately he tensed, whiskers twitching. He was about to rush out and confront whatever commotion was out there in his territory.

But as he turned the ground shook, and the tunnel caved in along half its length. His fight-or-flight impulses were thwarted by the fact that he was now buried in his own den in pitch-darkness.

The ground around him continued to quake, under pressure from above. The solid roof over his head—the underside of a log—groaned and shuddered. Then, through a landslide of earth, a streak of bright daylight flooded past one edge of the trunk. Finally, with a rush of falling dust and breaking sticks, the log lifted up and flew away.

The squirrel was out through the gap the instant daylight appeared. A huge clawed foot smashed the ground just behind him. The spot where he had been resting seconds before was now a crater two inches deep. The depression was in the shape of a bear paw, wider than he was long. The squirrel dodged through an avalanche of pine cones and dashed to open ground. He spun around and found a massive bear standing over his den, looking to dig up chickarees.

The squirrel ran back, so close to the bear that he broadcast his staccato chirps and buzzes straight up at the mountain of fur. He bounced three meters away when the bruin stepped over a fallen trunk, then he rushed right back again, never stopping his harangue. He circled the ponderous beast with angry chatter, but the intruder took no notice.

When the squirrel appeared, the bear had ceased her digging. Chickarees were impossible for her to catch when they were not confined in tunnels. The bear shifted her attention to the cones she had trampled. The squirrel kept up his rant until all the forest knew what was going on, but to no avail. The bear was an immovable object, parked on top of his home, methodically chewing on his stores of nuts.

The squirrel exhausted himself, and his scolding trailed away. He looked around, past the oblivious intruder, and realized he was out in the open. And he had announced his presence to any hawk or stoat within earshot. He grew quiet, then backed off and sought cover.

WITH THE AFTERNOON, the sky shaded to overcast. Chill gusts of wind raced up through the canyon to scatter leaves across the hillside. The squirrel returned to his den, but the site was unrecognizable. He identified his dead stump by its scent, even though it had been reduced to shards half buried in plowed ground. Detached pine cone scales littered the area. He stood on his haunches and listened through the stiffening breeze.

Nothing remained of his pile of cones and his cache of nuts. The scent of bear hung everywhere. In minutes the big looter had devoured the larder the squirrel had spent the summer stocking. The bear was provisioning her own upcoming hibernation, at his expense.

He searched the ground, poking his nose down into the soil. The acorns, chestnuts, walnuts were all gone. After a while he found a few stray pine nuts, which he ate. Then he set off to find a more secure site for another den.

He would have to find a lot of food in a little time to last him through the dark days ahead, but there wasn't much left to be found. He had already scoured his territory, uncovering seeds lodged in the most unlikely places. Elsewhere in the forest, other squirrels had staked out the most fertile territories—and they would defend their own nuts beyond the point of drawing blood.

But all he knew was how to gather and stash. So he busied himself again with the search for provisions, scouring the forest floor for nuts, looking where he had already looked many times before. He paused again to test the wind. It was already carrying the scent of rain.

A MONTH LATER, the site of the bear dig was a muddy patch on the forest floor. A succession of storms had passed through the area, and the ground was saturated. Below the surface, one lost acorn was soaking in rainwater.

Despite the tendency of acorns to fall straight down, like a rock, that acorn was now in a place higher than the treetop it came from. This it had accomplished by being delivered to the new site in the jaws of a chickaree. The squirrel had found the strength to carry the acorn upslope by eating other acorns. The mother oak tree had provided that energy. This enabled the busy squirrel to move some of the tree's seeds out of her shadow. A few of those would eventually come to be planted in good growing spots.

This particular acorn had rested in the rocks for years. Now it was coming to life in plowed soil. A matrix of dried proteins packed in crystalline storage within it had absorbed enough water to break the seed's dormancy. A switch had been thrown. As they swelled with rain, the oak cells in the acorn were coming to life.

The germination switch synchronizes what happens inside the acorn with what happens outside. The switch detects the color of the light filtering down through the soil. The acorn would not germinate too deeply buried; neither would it germinate exposed on the surface, where it would dry out in the sunshine; it would not even germinate in deep shade. Successful seeds send up their shoots only from spots that are unusually wet but also well drained and lighted.

The seedlings succeed because the germination switch senses the reliability of the water supply. A seed can swell with rainwater, but if that water does not linger the seed will not germinate. Instead, it will dry out again, shrink back to dormant size, and wait for better conditions.

The germination switch predicts the springtime suitability of an acorn's situation based on what happens in the winter. This ability to coordinate germination with environmental cues has been honed over millions of years of selective evolution. The big seeds will rest in stasis for decades, waiting for the ground to shift around them and provide better conditions. Many of them die while waiting.

Most of the weight of an acorn is starch, a source of biological energy. Most of the mother tree's investment in seed starch serves to feed squirrels— and jays, and acorn woodpeckers, and any other of the animals that move oak seeds around through the forest. But in the few seeds that survive, seed starch serves to power the germination of seedlings. In this buried acorn, the starch was already being broken down into sugar. The sugar was being built back into the walls of expanding, multiplying cells.

A few weeks after the acorn had broken out of years of dormancy, its tip swelled and a taproot broke through and bent downward into the soil. By the time the winter began to thaw, a green, growing oak shoot had emerged into the light, with a taproot beneath it two feet long. The root continued to grow down, staying level with the receding water table. It would support the growing shoot as it expanded skyward. By summer, a sapling oak was well established over the spot of the bear dig.

FIFTEEN YEARS LATER the oak tree had grown wide and tall enough to cast its own trunk in deep shade. The tree had been growing faster every year. The ground at its base retained no trace of the circumstances of the founding acorn's germination.

This season the tree would pause and shift its growth energy to the production of its own acorns. For years now, it had flowered and set seed. But each year it had aborted those fruits long before they matured. The tree had remained focused on its own expansive growth. This year would be different. Another switch had been thrown, and the tree's energy would now be diverted to filling all of its seeds with starch.

This switch functioned in counting the seasons. It would commit the oak to total seed set after a minimum number of years had passed since it germinated, or since its last heavy seed set. And, like the acorn's germination switch, the seed-set switch was attuned to environmental conditions. It sensed a series of cues—a late-winter frost followed by an unusually wet spring, a lower-than-average incidence of oak gall wasp parasitism—indicators that were broad enough that they would also be noticed by the other oak trees across the area.

The seed-ripening switch coordinates what happens inside the oak tree with what is happening outside of it. All the oaks in this forest carried the same switch. They were all counting the years since the last big seed set; they had all felt the same environmental cues. This meant that they would all be switching their energies to seed production at the same time.

The synchronization of their seed production is another facet of oak survival developed over millions of years of evolution. Such masting helps the oak trees manage predation on their acorns. This particular fall would bring a surprisingly heavy rain of them. Acorns would be carried all over the forest by all sorts of animals, and more of them than usual would be misplaced or forgotten—and left to germinate in the winter.

More acorns would be available this year than all of the animals that customarily eat acorns would know what to do with. Squirrel larders would feature acorns prominently, and squirrel survival over the winter would be better than average. It would also be a good year for bears.

REFERENCES

Goheen, J. R., & R. K. Swihart. 2003. Food hoarding behavior of gray squirrels and North American red squirrels in the central hardwoods region. *American Midland Naturalist* 158, 403–14.

— 12 —

The Sweetest Niche

THE FLOWERING PLANTS are a hallmark of life on Earth. The other terrestrial organisms—the fungi, the animals—depend upon them for their sustenance. Those other organisms can reproduce independently. In contrast, many of the flowering plants are not so self-sufficient. The work of spreading their pollen from flower to flower has been outsourced to the animals. The production of the seed through which these plants reproduce is dependent upon this alliance with members of another kingdom.

The distance between two flowers on a blooming bush seems miles long—when measured on the scale of a microscopic pollen grain. The pollinators bridge that distance. The plants supply them with nectar, which fuels their flight between blooms. Nectar sustains the mutual relationship. It links the flowering plants with their pollinators.

Three of Earth's kingdoms of life come together around nectar: the plants, the animals, and the microbes. Nectar is a complex, vital solution. It contains enzymes and small molecules of flavor and aroma. It contains amino acid molecules, as well as antibiotics like hydrogen peroxide. It also contains colonies of living microbial cells.

The essential microbes in nectar are the flower yeasts. A flower provides a unique niche for them—the yeasts depend on the nutrients in nectar for their energy and their growth. They depend on the plant's pollinators for their mobility between flowers. In return, the yeasts work to ensure the vitality of their niche.

There are more than a thousand different kinds of yeasts (and these are just the ones we know of). They are adapted to occupy the most eclectic of niches, from the surfaces of roots and leaves to the skins of fruits and the tongues of bees. The particular yeasts adapted to life in nectar have a distinctive shape. They are rods that grow in short clusters, attached through their bases—giving them the appearance of irregular snowflakes drifting in their special solution. They live suffused in the pastel light that passes in through a flower blossom's walls. Their tiny propeller shapes, rotating slowly, serve to preserve the nectar they live in.

The flowers need to keep their nectar unadulterated for the sake of their pollinators, who survive by drinking it. The flowers also need their nectar to be pure for their own use. They resorb what remains of it back into their own veins after pollination has occurred. The flower yeasts sustain the purity of that nectar.

The flowers live in an environment that is full of microbes. There are bacteria that live on the leaves, and filaments of fungus that live in the dirt and float on the dust. Those organisms cannot grow in solutions as high in sugar as nectar is—25 percent or more. But the yeasts can tolerate high sugar levels, and so can the mold fungi.

Millions of mold spores share the environment of each flower. But molds require amino acid molecules in their environment, from which they build their proteins. The flower yeasts have a very high affinity for those molecules. With yeasts established in the flower's nectaries, the mold spores cannot germinate for lack of amino acids.

Their suppression of competing microbes allows the yeasts to grow to high concentrations in nectar. If molds or bacteria grew to those concentrations, they would spoil the sweet drink, and pollinators would be repelled. But the flower yeasts are benign. Even when they reach concentrations that cloud their solution, they do not dissuade the flower's visitors. These yeasts even flavor their nectar with particular scents that enhance its attractiveness to pollinators.

THE MUTUALISM OF the plants and their pollinators has been perfected over the eons. The flowers communicate with the bees by sight and by scent, and by other means as well. Overnight, the blossoms build up a faint electric charge that extends out to the tips of their petals. When a bee approaches such a flower in the morning, the charge reaches out across space and raises the hairs on her body. This tells the bee that this flower has not yet been visited today—it should still be filled with a few microliters of nectar.

Should the bee or another insect visit the flower again later in the day, even before touch-down it will notice that the charge on this flower has been dissipated by a previous visitor. Then the pollinator won't even bother to alight—most of the nectar and pollen will already have been harvested.

So, first thing in the morning, most of the flower yeasts disappear from the flowers—taken away incidentally by the foragers that come to sip up their energy drink. After that, only a trace of the sweet liquid and its yeast population remains. The cells cling in a thin sheet to the walls of the floral vessel. But that is enough to continue the cycle as long as the blossom survives. The yeast multiply again overnight, while the plants refill and replenish their nectar and restore the charge on their petals.

THE NECTAR YEASTS disperse among the blossoms the same way the pollen grains do. They both act as passive particles, dependent upon random chance for the completion of their journeys. Both types of propagules are produced by the thousands in each flower. Each one of them has the capacity all by itself to propagate its species through the generations. Ninety-nine percent of them will fail, consumed by the pollinators or lost somewhere in spring's flowerscape. But they are produced in such excess that some of them are likely to succeed and carry on their lineage.

As the pollinators pursue their nectar, they pick up the pollen that is positioned close to each nectary. They then spread those grains among all of the blooms of each flower they visit. At the same time, the pollinators spread the flower yeasts far and wide, to all of those same flowers.

Each yeast cell, and every pollen grain, carries an entire genome at its core—a few nanograms of DNA. These genomes encode instructions that guide the growth of an entire colony of yeast cells, or an entire plant. The yeasts that chance to land in the nectar of an uncolonized flower will divide again and again, to establish a population of cells that will stabilize that nectar. Pollen grains landing on the stigma of a flower of their own species will also grow. There they begin the cycle of seed production and plant reproduction, assuring the continuity of summer's future flowers.

As September evolves toward autumn, on days when the insects are too chilled to fly, the nectar in the flowers might simply accumulate and only grow yeasts. Instead, the flowers simply shut off nectar production when it gets cold. Finally, at the end of the growing season, the flowers senesce. The plants shift from producing more buds or more nectar to producing seeds. The old blooms die and fall, along with their yeasts, to the ground, beyond the reach of their transporters—the cycle is broken. The flower yeasts must then survive a flowerless winter. Their success at hibernation will determine their success at reinitiating their cycle next spring.

NECTAR IS CONVERTED by the bees into another complex, living solution: honey. The bees maintain the vitality of the honey in the hive, where it sustains them and their young. They continually adjust its content to less than 20 percent water. They put the excess that they did not eat or feed to their young back into the honeycomb.

When bees sample their honey, they add the enzyme diastase to it. Diastase continually builds up or breaks down sugar polymers, which alters the viscosity of the solution. Diastase keeps the amber drink in its liquid form. It remains drinkable, even though honey contains more sugar than water can normally dissolve without solid sugar crystals growing through it.

Honey is a supercooled fluid, never solidifying even when temperatures are below freezing in the winter. But if it is removed from the hive, it dies. Then it solidifies, darkens, and eventually grows mold on its surface. Dead solid honey will not melt even on the hottest summer days.

Honey in the beehive contains everything that was in nectar: antioxidants and antibiotics, amino acids, enzymes, and live flower yeast cells. The yeasts cannot grow in living honey because of the low water concentration, but they can survive in it. While the honey steadily sucks the water from their bodies, the yeasts rest suspended in the solution and wait for conditions around them to improve.

The bumblebees are native pollinator bees in the New World, and in the temperate reaches of the Old. The pollen they gather is used as food for their own next generations. Larval bumblebees treat the plant's germ-line pollen cells simply as sources of nutrition. The information stored in the pollen cells' DNA is destroyed as the larvae break it down and extract its phosphorous to build their own cells.

Annual flowers emerge in temperate spring, die in the fall, and disintegrate into the ground before the winter. Their loss destroys the warm-season living space of the flower yeasts. Most of the insects that visit these flowers are also annuals. They appear in the spring, grow in the summer, and die back in the fall, just like the annual flowers. Those insects spend the winter as the next generation's eggs, while the annual flowers spend it as seeds.

But some insects ride out the winter in adult form. One of these is the queen bumblebee. When spring arrives, she emerges from hibernation to build a new underground hive chamber from scratch. There she lays her first batch of eggs. (She was already fertile, having mated the previous fall with a drone (or drones) who did not themselves survive the winter.)

Bumblebees are some of the first insects to become active early in the spring. They are warm-blooded animals. The foundress queen broods her eggs well above springtime temperatures (at 30°C). She is black—the color best suited to absorb the heat from morning sunlight that slants through grass still wet and cold.

The foundress sets out to forage as soon as she can generate enough heat to fly. From the first flowers of spring, she harvests nectar and pollen and feeds them to her nascent first brood. She will raise that brood of worker bees by herself. Her larvae mature and spin their cocoons, then around May Day, they hatch to populate the hive. After that, the queen never again leaves. Her first-generation progeny will support her in place all summer. They take over the tasks of foraging, expanding their burrow, and raising her successive broods, until the hive has grown to contain more than a hundred bumblebees.

The worker bumblebees construct wax vessels and fill them with honey, which they synthesize from nectar. Toward the end of the summer, the hive produces drone males and new reproductive females. In the fall, after her year is over, the foundress queen dies, along with all her workers and the drones.

The new reproductive females fill their crops with the remainder of the summer's honey. Then they leave their nest and set off to find hibernation spots and to continue the cycle. The abandoned hive disintegrates in the rain and is consumed by fungi and swarms of tiny soil arthropods.

The hibernating queen bees are vessels. They carry the eggs within them that will perpetuate their species into the future. They also carry honey in their crops, to sustain them through the winter. The flower yeasts in that honey are carried over the winter inside the queen bumblebee.

Soon after the queen emerges into spring, her tongue introduces the flower yeasts into the nectaries of the first flowers. The yeasts begin to colonize those blossoms, reinitiating their cycle once again. They multiply and wait for other insects and birds to spread them to the rest of the flowers. Subsequent flower yeast generations will be borne by other pollinators to populate each of the blooms that appear over the successive months. The flower yeasts will stabilize the nectar within them all.

In this way, the flower yeasts switch back and forth between plant and animal hosts. They divide their time equally between summer, when they grow in the translucent, pastel nectaries, and winter, when they rest in darkness, in the honey inside queen bees. Through all the switching, the yeasts remain immersed in the sweetest of niches. They live year round with their sustenance guaranteed and with their transportation between their two alternative living spaces reliably provided for them.

NOTES

The yeasts that live in nectar are in genus *Metschnikowia* (Herrera et al., 2009) which is classed with the Ascomycetes, along with budding baker's yeast. Nectar provides their essential amino acids for growth. The flower yeasts have a very high affinity for free amino acids, and they may prevent the growth of molds in nectar by sequestering those essential molecules (Dhami et al., 2016). The pollinators depend on the sugar in nectar for their energy. The lack of available building blocks of protein in nectar is of little concern to the pollinators because they supplement their liquid diets with high-protein snacks: insect pollinators eat pollen to provide their protein, while the nectar-foraging birds eat the occasional insect.

The bees raise their larvae on a high-protein diet derived from pollen. The flower yeasts may reach high concentrations in nectar, but their presence is not dissuasive to pollinators, as comparable concentrations of spoilage organisms would be (Good et al., 2016). Flower yeasts can be observed clouding the sugar solutions in hummingbird feeders, transported by the birds from flowers. The hummingbirds continue feeding at the clouded feeders. That suggests that the flower yeasts do not inhibit pollinators (Good et al., 2016) even at high concentrations.

Nectar contains superoxide dismutase proteins that synthesize hydrogen peroxide. It also contains ascorbate reductase, which keeps ascorbic acid in the form that scavenges the damaging free radicals produced by peroxide. And it contains carbonic anhydrase, which buffers the nectar, keeping the acidity of the solution in the normal range, so protecting against the tendency of peroxide solutions to become alkaline (Carter & Thornburg, 2004). (Blood-borne carbonic anhydrase performs the same buffering reaction in the veins of the animals.) Yeast cells contain peroxidases that protect them from peroxides, which are toxic to bacteria. Metabolic products of flower yeast culture may enhance the attractiveness of nectar (Reiring et al., 2017).

Other yeasts closely related to the flower yeasts live in lower-sugar environments. They compete for their niche space by producing molecules that are toxic to the bacteria and fungi, such as collagenase and iron-sequestering siderophores. The flower yeasts have not been found to produce such antibiotics. (Those toxins might poison the microbes that live in symbiosis within the pollinators that feed on their nectar.) Yeasts tolerate high-osmolarity (low water activity, high sugar concentration) solutions. Each sugar molecule directly binds about twice its weight in water molecules, structuring the water around it. Other water molecules associate peripherally with that structure. That water structure inhibits the growth of microbes adapted to live in (unstructured) water.

Many flowering plants (oaks, pines, grasses) are wind-pollinated. The flowers that are insect-pollinated communicate with their pollinators by color and scent signals, as well as by static electric fields (Sutton et al., 2016). (European honeybees are not native to North America; they were introduced in the 1700s.) The pollinators have established a mutualism with the plants that is shared by the flower yeasts. The yeasts are spread among summer's blooms by pollinators and are sustained through the winter inside bumblebees (Pozo et al., 2018). Both sides of their arrangement cooperate passively; neither the yeast nor the pollinator is physically taxed by or physiologically dependent upon the other member of the pair. Their relationship simply involves transport, a form of mutualism called phoresis (the yeast member of the arrangement is called a phoront).

REFERENCES

Carter, C., & R. Thornburg. 2004. Is the nectar redox cycle a floral defense against microbial attack? *Trends in Plant Science* 9, 320–24.

Dhami, M., et al. 2016. Genetic basis of priority effects: Insights from nectar yeasts. *Proceedings of the Royal Society B* 283. doi: 10.1098/rspb.2016.1455

Good, A. P., et al. 2014. Honey bees avoid nectar colonized by three bacterial species, but not by a yeast species isolated from bee gut. *PLoS One* 9. doi: 10.1371/journal.pone.0086494

Herrera, C. M., et al. 2009. Yeasts in floral nectar: A quantitative survey. *Annals of Botany* 103, 1415–23.

Pozo, M. I., et al. 2018. Surviving in the absence of flowers: Do nectar yeasts rely on over-wintering bumblebee queens to complete their annual life cycle? *FEMS Microbiology Ecology* 94, fiy 196.

Reiring, C. C., et al. 2017. Nectar-inhabiting microorganisms influence nectar volatile composition and attractiveness to a generalist pollinator. *New Phytologist* 220. doi: 10.1111/nph.14809

Sutton, G., D. Clarke, E. Morley, & D. Robert. 2016. Mechanosensory hairs in bumblebees (*Bombus terrestirs*) detect weak electric fields. *Proceedings of the National Academy of Sciences* 113. doi: 10.1073/pnas.1601624113

— 13 —

Drifters

THE MIDREACHES OF the South Seas are months of drifting-time away from any shore. The southern hemisphere of this planet is 90 percent ocean—the seascape there is wider than any desert. It is barren in every direction, silent all the way around the far horizon. By day the surface is becalmed under relentless sunlight; by night it chills to mirror the stars of Capricorn and the Southern Cross. Terra firma is not part of the story here. The solid bottom is thousands of feet below, steeped in perpetual darkness.

These desolate reaches do not support the great numbers of vertebrate creatures that abound in the more productive polar waters to the south. None of the residents of this desert swim with great speed or fly for great distances. This is the province of the low-energy, simple drifters.

They are invertebrates, jelly creatures. Usually they cannot be seen. A solitary blue button may come into view drifting ten feet down. Only the disk is apparent, and it soon fades into the blue background—returning the seascape to its desolation.

However, this realm is capable of flashes of great productivity. The chance convergence of air and sea currents can animate a sea that had previously been becalmed for months. Sunrise may reveal a surface littered as far as the eye can see with small sails. Those would be by-the-wind sailors. Millions of them per acre appear overnight. They arise over an upwelling from the depths, which produced an explosion in the numbers of their invisible planktonic prey.

The sailor's small rafts are only inches in diameter. They appear to have been assembled from three half-circles of cellophane. Two are fused into an oval that sits flat on the water; short tentacles beneath the disk capture microscopic plankton. The third half-circle is a top-side sail, standing high and dry. Should the creature be flipped by a white-cap, the heavier tentacles roll back below the lighter sail to right the ship.

Like the blue button, the by-the-wind sailor's upper sides are deeply colored. This is defensive counter-shading: these animals blend in with the background of the deep dark sea when preying eyes pass above. And from the other direction, looking up, they are transparent enough for the daylight to shine through, rendering them almost invisible against the sky.

The sailors get their nutrients when their micro-venomous tentacles capture their microscopic prey. But they get their energy from the sun. They live bathed in bright light all day long. Thus they are well situated to grow a garden in their transparent tissues. They are part plant: they carry internal photosynthetic algae. The algae donate the sugars they produce to the vessel they ride in. Those single-celled symbionts are much like the free-living plankton that live in these waters. But as passengers within their glassine ship, they are protected from most of the sea's many planktivores.

By-the-wind sailors are rudderless, completely passive. Their tentacles do not reach beyond the edge of their disks, nor down far enough to retard their rotation. The invisible fingers of the west wind trace wakes across the surface that spin the little vessels around. All the short sails rotate in unison, like compass needles magnetized by the gusts. In this orientation, sideways to the breeze, they are best set to catch a push-off toward the far horizon.

Then the flotilla of sailors heads east to new pastures. The journey will end for many with mass strandings. They are beached on the continents that lie astride the direction of the prevailing westerlies. On the sand, their bodies will dry paper-thin. They pile up in purple stripes across miles of high-tide lines on the beaches of Africa and the Americas.

AT NIGHT THE open ocean shows more activity: creatures that retire from the daylight return to the surface. Some swimmers appear among the drifters. Some of these are shellfish, including the sea butterfly. In contrast to the butterflies of the sky, this butterfly does not show off bright wings in the highlights of day. Like a moth, this creature knows only the nighttime. Before dawn rises it dives away into the depths. The increasing pressures below change the flavors in the water and offer greater resistance to movement, but the animal adjusts. Hundreds of meters down, the perpetual darkness keeps the butterfly hidden from visual predators.

This creature is another invertebrate, a mollusk. At depth, the plankton that sustain it cannot grow. Those microscopic drifters do not exist in darkness, where these winged hunters spend their days. But when the last traces of sunset fade above, the butterflies begin their ascent. If the light brightens as they rise, they pause and give the sun time to sink farther below the horizon. If they see shadows moving through the water above, they settle back down. They spend half of their lifetime pursuing their migrations up and back. When the last of the dusk has given way to stars from west to east, the sea butterflies appear in the midst of the floating plankton to graze on the surface.

The sea butterflies join a menagerie of other open-water creatures in this vertical migration. Lower members of the food chain from all the aquatic and marine phyla—from the lakes to the open ocean—retreat from the light like this every day. They avoid the diurnal vertebrate predators—the diving birds, the dolphins, the sharks, the sailfish—that may share their waters. Their daily trek up and down through liquid habitats everywhere is the largest migration in the world in terms of numbers and masses of moving participants.

The sea butterflies don't show themselves in the light, and the light does not show them either. They are transparent—light passes through, revealing only their internal organs. They are snails with glassine shells, but they are not bottom dwellers. They live in the open ocean, never treading solid ground.

They live upside down, with their transparent foot above them. That foot has evolved into two wings. In the largest members of their species, the wingspan is as wide as a swallowtail butterfly. They flutter through the water like a bat through the night. Then, as first light begins to rise, they fold their wings together above them and sink, shell first.

Where they pause to feed on the surface at night, the sea butterflies have adapted the strategy of the spider or the jellyfish. They spin a transparent web of mucus that extends out ten times as wide as their shell. The web trails off into the water and traps small floating planktonic creatures. Then the butterfly consumes its own web, along with what it has caught.

Despite their retiring habits, sea butterflies are hunted day and night. They live low on the food chain. When their populations bloom into swarms of hundreds of thousands in colder waters, they become prey for deep-diving baleen whales. But their greatest nemeses are members of their own kind. Sea angels, members of another pelagic molluscan order that have no shells at all, feed exclusively on sea butterflies.

ON ANOTHER DAY the surface is again calm and deserted as first light pools on the eastern horizon. Then a greenish column materializes, rising from the depths. It is lit by luminescent plankton that have proliferated in the water in recent weeks. They brighten when disturbed, and now they are being pushed aside by a rising creature. They illuminate its vertical wake.

The top of the column is a balloon that inflated itself below. It is accelerating upward with the buoyancy of a cork. It is the float of a Portuguese man o' war. The creature was resting in the depths, and now it has inflated itself. The water pressure lessens as it ascends, letting its float expand and tighten.

When the man o' war breaks the surface, the sail that frills its crest spins the creature sideways to the breeze. Its movement across the water is slowed, however, by the sea anchor of tentacles it has extended beneath its float. The whole creature is bathed in its own cold luminescence, which shines through its translucent air bladder. Its tentacles below are decorated with tiny sparkles of the green plankton it rose through, glowing at maximum brightness in response to a deadly sting.

The illuminated spectacle attracts attention. One of the largest creatures of this realm materializes out of the gloom and heads for the man o' war. It is an adult female loggerhead turtle, seven feet long, black, and a thousand pounds in weight on land (though weightless in the water). It homes in on the lighted beacon and attacks, devouring the whole creature except for a few of the glowing tentacles, which diffuse away on the current. As they drift downward, the fading strands will capture more of the hapless planktonic creatures, now to no purpose.

At dawn, more of the men o' war pop up all across the surface. The vivid pinks and purples of their inflated bodies are brightened by the color in the sunrise. A few small clownfish appear from nowhere to shelter in the shade of the floats—they are impervious to the stinging tentacles. The trailing strands become thinner and more transparent as they stretch a hundred feet down, to where the light has faded away. At depth, they paralyze and hold, and eventually haul up, whatever free-living things have the bad luck to brush against an invisible thread.

A calm-slick appears on the surface along the line where colliding ocean currents meet head-on and plunge straight down side by side. The men o' war accumulate along this path. Their flotation keeps them from being sucked downward. They do not pile up together in these calm slicks because they track along their courses at different speeds. Some of them haul up their tentacles, and so are pulled farther by the wind; others leave their strands extended, fishing through the passing flow, and the drag of the current becomes the main influence on their course.

Those ocean currents flow in closed loops. Far beyond the places where they dive toward the bottom, they rise back up again. Such upwellings are the hunting grounds in which these passive jelly creatures multiply. The rising currents bring up nutrients from the bottom. These nutrient-rich columns bloom with the microbes that anchor the planktonic food chain upon which the jelly creatures depend.

THE PREVALENCE OF the hydrozoan jelly creatures reflects their low standing in the open-ocean food chain. The solid-bodied members of the next rung up that chain are encountered less often. But where the flotilla of men o' war passes, some of the floats are partially collapsed, keeled over in the water. A few are so deflated as to be awash, near sinking. They are under attack from another of their predators, another open-ocean shellfish.

One of the men o' war appears to be slumped over onto a cushion of bubble-wrap. On the inner edge of that sheet of bubbles hangs an indigo-colored spiral cone. This is the violet snail. Its white tip points straight down, toward the rocks and silt far below, where shellfish are usually found. But this snail never sinks to the solid substrate; it lives on liquid, along with its prey. Its stone weight is suspended under a raft it has blown of durable bubbles that keep it afloat on the open water. When it encounters a man o' war, it docks with it and begins a snail's-pace attack. It is immune to the stings of the jelly creature, which it may take days to devour.

The convoy of men o' war is under siege on its far flank as well. There, the predators are sea swallows, members of the nudibranch family. They look like detached fronds from a purple fern floating on the surface tension of the water. They too are mollusks, but they shed their shells soon after they hatch from eggs. Several of them will attack a single man o' war and remain until it is consumed.

Most nudibranchs live on solid reefs and shoals. They are colorful animals that carry their gills exposed on their backs. Many of them feed on sessile marine flowers, the anemones, which are relatives of the floating jelly creatures. Those nudibranchs' gills are colored like flowers—they may be boldly yellow, or sky blue grading to orange. But their bright petals carry a message of warning. The color advertises the nudibranch's ability to consume the stinging cells of the anemone and then pass those cells undamaged, still alive and undischarged, through its body. Those stinging cells are finally emplaced among the gills on its back, like the spines of a nettle, facing upward—ready to sting.

A man o' war is an inverted anemone, anchored to the ceiling of the sea by its float, with its tentacles pointing not upward, but down. The sea swallow has conformed to this inverted order. It is an upside-down nudibranch, with its foot uppermost—walking on the underside of the surface, looking for jelly creatures as prey.

Like the nudibranchs, the sea swallow has retained the capacity to consume stinging cells and pass them through its system to defensive positions where they are maintained alive and armed. The body of the sea swallow is centered between projecting wings that look and act like the ray-fins of the lionfish. This nudibranch loads the tips of those projections with the most potent of the stinging cells of its prey—its sting is worse than that of the man o' war.

But the sea swallow, unlike the lionfish or the coral-feeding nudibranchs, does not wear bright warning colors. It wears the open-ocean defensive counter-shading: deep blue above, silvery-gray below. It, too, is far from the top of the food chain.

THE HYDROZOAN JELLY creatures are not all members of the surface community. Most of them live below the surface. This includes the deep-water siphonophores, the longest of the hydrozoans. These thin, sinuous colonies of polyps can be fifty feet long or more. Their transparency makes it difficult to see all of their length in the same field—until the night descends and they are outlined in their own luminescence. Some of their luminescent tips glow red and jiggle, attracting prey toward the stinging cells.

The individual hydrozoan polyps are joined end to end in these sipho-nophores. The long strand coils and folds back on itself with the current, until the advance of the swimming bell at one end pulls the chain straight again. Each individual polyp trails a single tentacle out into the current.

The siphonophores are covered in potent stingers. Nonetheless, as with the jelly creatures of the sunlit water, they are still attacked by mollus-can predators—free-swimming nudibranchs and sea angels—which search them out up and down the water column.

A new dawn finds the sea deserted and becalmed once again. From a vantage point just above the surface, the mirror image of the sun's disk appears to lie just a few feet away. It is angled as far below the horizon as its source—millions of miles away—is angled above it. Streaks of cirrus spread low along the western horizon. The ocean lies flat, anticipating stormy weather.

Off to the northeast, the blue sky and its reflection on the sea grade to shades so similar that the dividing line between them disappears. No birds dip and glide in the middle distance; none of their cries carry across the water.

To visitors from the land, the surface appears vacant and monotonous. The converging and diverging currents can generate desert-like emptiness over much of the seascape. But those same currents generate high vitality in other places. The variation provides a geography of different textures, temperatures, and scents.

The vast oceanic plain is divided by these differences. Their signs are recognized and followed by all the residents living at the interface of air and water. The sea birds—shearwaters, petrels, albatross—appear at places where their smaller prey have multiplied. The great Leatherbacks do the same below, as do the even bigger mid-ocean sunfish searching the dappled depths for aggregations of jellyfish. But most of the time, the vertebrate predators are far away, leaving the tropical southern ocean to the drifters.

NOTES

The true jellyfish are individual creatures, but the jellyfish-like hydrozoans are colonial—made of masses of small, individual, independent hydroid polyps attached to each other. In the siphonophores, they are attached in a linear row. In other hydrozoans, differentiation among members of the polyp colony produces jellyfish-like shapes, e.g. in the blue button (*Porpita*), the man o' war (*Physalia*), and the by-the-wind sailor (*Vellela*).

The by-the-wind sailor's sting is lethal to microscopic organisms, but you would not feel it on your hand (though it could sting your eye if you rub it after handling *Vellela*). The sails of *Vellela* orient perpendicular to the wind because a freely rotating airfoil assumes the orientation of greatest resistance to an airflow. The sailors strand on beaches in Chile, Australia, and other west-facing continental shores.

Their larger hydrozoan cousins, the men o' war, strand less frequently. Their courses are determined more by the current than by the wind—by near-shore currents that parallel the beach, as opposed to the sea breeze (which propels *Vellela* into the sand). Men o' war can sink into the water,

avoiding surface effects and predators, then they can inflate their gas bladder, and quickly rise (Whittenberg, 1960). To return to the surface, they
generate carbon monoxide gas (apparently they are resistant to CO poison-
ing), but the other gasses of the air diffuse into their float once it reaches
the surface.

The hydrozoa and the other jellyfish are in the phylum Cnidaria. They
are all preyed upon by members of the phylum Mollusca, which include
gastropods in which the foot has been transformed into a wing (Pteropods;
Hunt et al., 2008). The sea swallow, a free-swimming nudibranch (pro-
nounced "nudibrank") is a member of that group. The nudibranchs can
manipulate the living stinging cells (nematocysts) of their prey, moving
them intact through their bodies to defensive external placements (Greenwood, 2009). The sea angel (*Cephalopyge trematoides*) is a free-swimming
pteropod molluscan predator of subsurface siphonophore hydrozoans.

REFERENCES

Greenwood, P. G. 2009. Acquisition and use of nematocysts by Cnidarian predators. *Tax-
icon* 54, 1065–70.
Hunt, B. D. V., et al. 2008. Pteropods in southern ocean ecosystems. *Progress in Oceanography* 78, 193–211.
Whittenberg, J. 1960. The source of carbon monoxide in the float of the Portuguese man-
of-war. *Journal of Experimental Biology* 37, 698–705.

— 14 —

Lace Lichen

AIR PLANTS HAVE forsaken the soil—they get what they need from the sky. In the tropics these epiphytes bloom in profusion hundreds of feet up in the trees. They survive with no connection to the earth. They intercept their rainwater before it gets to the ground. Their roots serve mostly to hold them on high.

But far from the rain forest, air plants still thrive—in places where arboreal orchids, ferns, and bromeliads would wither. These temperate epiphytes have no roots at all; they are nurtured solely by the air. They don't depend on the rain to supply the water for their respiration—they live where months can pass without precipitation. They absorb their water in its invisible form, as vapor.

The temperate epiphytes fill air-plant niches in the conifer and hardwood forests of coastal New Zealand and Chile, and on the west coasts of Japan and Northern Europe. The springtime air is often dry by noon there; for long stretches of the summer, it is dry all day.

Those woods reach their greatest extent along the eastern rim of the North Pacific. They occupy a coastal arc stretching from Alaska down through the inland mountains that cross into Baja California. Gauzy, weightless, gray-green curtains dominate the air-plant niche there.

These air plants are lichens—associations between fungi and algae. Where they grow alone, each organism produces only a thin coating on the bark. But when they unite, they realize the benefits of synergy. They lengthen into one of the hallmarks of the northeast Pacific coastal forest: the hanging tapestry of lace lichens.

In contrast to bromeliads and vines, which absorb their water in liquid form, lace lichens are not succulent. They have no veins to carry water through stems and out to leaves. Each cell absorbs its own water and nutrients independently of its neighbors. The lichen's shape promotes this: no cell in its tracery is more than a tenth of a millimeter removed from thin air. The flattened, finely divided foliage provides these lichens with the greatest surface-to-volume ratio of any plant.

The structure that accommodates that geometry is a latticework of open ovals woven together from living fabric. The weightless webbing does not offer much shade or wind resistance. Its filaments hang limp and dusty gray when dry—twisting in the breeze like silk curtains billowing behind an open window.

The strands can be found trailing above yellowed grasslands in the interior hills, far from the more humid conditions near the ocean. There they wave like pennants, marking out a course through the maze of valleys in the mountains behind the shore. That is the course along which the sea breeze spills farthest east overnight. It carries the cool humid coastal clime inland.

Unlike plants whose roots tap into the reservoir of moisture in the earth, the lichens cannot run their photosynthesis all day long. Their inability to retain moisture in their flat filaments prevents them from growing continuously. Pores in the watertight plant cuticle must open to admit carbon dioxide gas. CO_2 is the source of the carbon the plants polymerize into fiber. But pores cannot control the direction of gas diffusion—it goes both ways—and water vapor escapes when the pores are open. Lichen does not raise water from the ground through roots—it depends on the air. When the air dries, a lichen wilts if it leaves its pores open. So the lace lichens shut down its pores, and photosynthesis pauses for lack of CO_2—just when the sun becomes bright enough to warm the morning.

Lace lichens shrink from the glare of a hot afternoon. They avoid full sunlight—bearding the shady lower scaffolds of their host trees. They eventually lose almost all of their water and spend their dry spells desiccated to the consistency of crêpe paper. In their driest sites they may show no signs of life for years. Yet they soften again in seconds at the first breath of fog or sign of rain. Their photosynthetic metabolism comes to life when they rehydrate; the revitalized gray-green latticework will stay soft and supple as long as traces of humidity linger.

THESE LICHENS LIVE the unattached life. They harvest all they need from the empty sky. They absorb their sulfur, carbon, and water directly, as gasses. Some lichens also absorb nitrogen gas from the air and then convert it into its biologically useful organic form.

The sky also carries nutrients that are not available as gasses. These come in the form of a microcosmic sampler of dusts from lands the world over. They include miniscule grains of sea salt. Those crystallized from droplets sprayed into the sky from breakers crashing on distant shores. The smallest of those droplets evaporate completely before they fall back into the surf. All that remains is fine, levitated particles of salt. Air-borne, they are entrained in the sea breeze and carried into the sky.

Other solids carried on the wind include fine particles first raised by sand storms in the Gobi Desert; volcanic aerosols injected into the jet stream above Indonesia; invisible smoke particles that rose in Asia—all of them end their journey in the finely divided lichen foliage.

The webwork of the lace lichens serves to becalm the air entrained within it. This air carries these motes from the farthest corners of the Earth. The microscopic particles wander into the living lattice and settle on the strands. Eventually they dissolve in place in predawn dew. They color a tea that coats the gray-green surface with a dilute solution of phosphates, sulfates, potash, and other trace minerals. All plants need these nutrients, but this one plant does not need to search for them in the ground.

THE LACE LICHEN's habitat is a land of two seasons—wet and dry; spring and summer—each lasting six months. Lace lichens are among the fastest-growing plants in this environment. They grow a third again as long over a season of spring mornings, sometimes reaching twenty feet in length. When the flowers below them are finishing their season, the lichens suspend their lives and wait out the dry heat in stasis. They accumulate a summer's load of dust, primed for the reappearance of the dew when the weather turns once again. Finally the rains come. They wash away the insoluble coating that remains of the sediment that settled upon them. The cleaning brightens the blue-green foliage to its maximal photosynthetic potential.

Where they grow the fastest—near the coast—the lacy lichens do not reproduce by making spores. They are spread from tree to tree through the temperate woods as fragments, often with the assistance of the animals. Just as other plants rely upon their flying allies to disseminate their pollen or seed, this plant spreads its population on the wing. Its partners in this dispersal are hummingbirds.

The hummers build their nests from strands of lace, having adopted lichen as a material of choice because of its particular qualities. They gather the tough, dry strings from mature stands of lichen, and weave them with other fibers into a tiny cup stitched with spiderweb. They line the inside with plant down and decorate the outside with bits of the wider flat ribs of lace lichen, and with other types of lichens as well.

The hummingbird nest is built to be expansible: it enlarges to accommodate the growth of its chicks. As the hatchlings mature and their short beaks begin to lengthen, their respiration hydrates the strands of lichen next to their bodies. The dry filaments come to life when they absorb water vapor from the chicks, just as they do from the sky on a dewy dawn. Water that was nectar in flowers at daybreak, and chick food in the morning, is absorbed by the lichens later in the day, allowing the strands to expand and become pliable. The nest stretches with the press of the growing hummers, dilating to make room. It may nearly double in diameter as the fledglings' plumage fills in and fluffs out over the weeks.

Other weavers also transport lichen strands for nest building: orioles, bushtits and gnatcatchers. Later, the empty nests hang vacant through the dog days of summer, sagging and drying out, having accomplished their first task of fledging the next generation of weaverbirds.

With the fall, the lace in the abandoned nests comes to life once more. The stringy strands absorb the season's moisture and begin to grow again, overflowing their woven confines. They trail down into the open air spaces below, expanding over each untended nest until it is unrecognizable. Through the years, all the best nesting sites come to be hung with lace. The open understory spaces grow curtained in a gray-green gown that trails all the way down to the browse line—where the deer trim the longest streamers.

LACE LICHENS SURVIVE in intimate association with the air. That lifestyle leaves them susceptible to recent atmospheric changes that can overload their metabolism and poison them. They have never evolved a mechanism to curb their accumulation of airborne minerals, which were always so scarce that every last molecule had to be absorbed if the lichen was to prosper. As a consequence, their capacity to take up all the trace gasses they touch will overdose a lacy lichen on sulfur dioxide, or oxides of nitrogen, if those gasses are present in excess. They flow into lichen strands in excess, from anthropogenic air pollution.

Lace lichens are no longer found in their historic range in the Los Angeles basin. The smog has killed them. Their growth there is now limited to slopes in the surrounding mountains at altitudes above the prevailing ceiling of the inversion layer. Some residents of southern California—those living north or south of the plume of pollution that stretches inland toward the desert—still find lace lichen in their trees. Its persistence is an indication that the local air quality at those spots is not harmful to the health of plants, or to anyone else sharing that air.

NOTES

Lace lichen, *Ramalina menziesii,* is sometimes called Spanish moss, but this would be a double misnomer. The Spanish moss in tropical ranges in the Americas is neither a lichen nor a moss, but a vascular, flowering plant (a bromeliad, the Latin name of which translates roughly as "looks like an epiphytic lichen"). *R. menziesii* is a temperate, nonvascular fungal vessel containing algae inside. The expansible hummingbird nest incorporates strands of lace lichen, along with spiderweb, which also lengthens when hydrated. Aside from propagation aided by hummingbirds (Larson, 1989), lace lichens also spreads by fragmentation. Along the coast, where their growth rate is highest (Boucher & Nash, 1900b), the strands do not sporulate (Larson, 1989). Where lacy lichen does form spores, they are only from the fungal partner of the symbiont; germinating spores must find a source of their algal partners before the strands can grow lace (Werth & Sork, 2008). Lichens absorb their nutriment without roots or contact with the soil (Nieboer et al., 1978); they can be a significant source of organic nitrogen for the forests that support them. The pendulous lichens are utterly dependent on what they absorb from the air (cited in Matthes-Sears et al., 1987), including dust, which they acquire by dry deposition (Boucher & Nash, 1990a). They assimilate airborne matter, so they are sensitive barometers of air pollution. They become dangerously radioactive when they grow

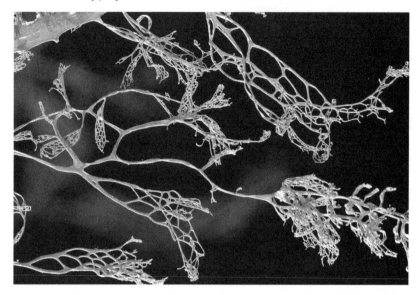

downwind of nuclear material contamination. Even before rock-bound lichens (which derive some of their essential elements from the rock) are extirpated from polluted regions, lace lichens disappear. They have retreated from the Los Angeles basin and are now found there only above the smog line (Sigal & Nash, 1983).

The lacy lichens cannot run the photosynthesis reaction and grow when the sun is brightest. This is because the lichens must close their pores to preclude wilt when the day warms up. Nevertheless, the hanging lichens can make up for sun-time lost in the afternoon. They do this by running the photosynthesis reaction at higher speed in the early light, before the water diminishes in the exposed cells. They can do that by raising their temperature against the cool of sunrise. The photosynthesis reactions run faster when the temperature is higher. The lacy lichens raise their temperature when they condense dew across their thin strands. Just as evaporation cools a surface, the reverse reaction—condensation—warms a surface. Evaporation involves a change in the state of water, from liquid to vapor. This process takes heat away from a surface, generating a chill that can be felt, for example, by moistening a patch of skin and then blowing across it. Reversing the process reverses the effect: heat is deposited into a surface when water in its vapor state changes back into liquid upon that surface—the process of condensation.

The fine divisions of the surfaces of lace lichens maximize the area available for condensation of the dawn vapors. They take up the heat provided during condensation of the dew upon their networks, and then they take up the dew itself (and everything dissolved in it). And, for a few hours each day, before the warmth of the approaching noon steals their water again, they actively grow.

REFERENCES

Boucher, V. L., & T. H. Nash, 3rd. 1990a. The role of the fruticose lichen *Ramalina menziesii* in the annual turnover of biomass and macronutrients in a blue oak woodland. *Botanical Gazette* 151, 114–18.

Boucher, V. L., & T. H. Nash, 3rd. 1990b. Growth patterns in *Ramalina menziesii* in California: Coastal versus inland populations. *Bryologist* 93, 295–302.

Larson, D. W. 1989. Some functional aspects of the net-like morphology of *Ramalina menziesii* Tayl. *Functional Ecology* 3, 63–72.

Matthes-Sears, U., et al. 1987. Ecology of *Ramalina menziesii* VI. Laboratory response of net CO_2 exchange to moisture, temperature, and light. *Canadian Journal of Botany* 65, 182–91.

Nieboer, E., et al. 1978. Mineral uptake and release by lichens: An overview. *Bryologist* 81, 226–46.

Sigal, L. L., & T. H. Nash, 3rd. 1983. Lichen communities in conifers in southern California mountains: An ecological survey relative to oxidant air pollution. *Ecology* 64: 1343–54.

Werth, S., & V. L. Sork. 2008. Local genetic structure in a North American epiphytic lichen, *Ramalina menziesii* (*Ramalinaceae*). *American Journal of Botany* 95: 568–76.

— 15 —

Swallow-Tailed Gull

THE EARLY BIRDS of the open ocean set out before sunrise. They take wing over their islands, heading for the far horizon. They are driven to bring back larger catches during the nesting season, while they are producing eggs. They will need even more when those eggs hatch into ever-hungrier chicks.

But when they arrive above their sunny hunting grounds they find only slim pickings. Most of the open-ocean prey species have disappeared before the birds appear.

A menagerie of animals animates the sea's surface all night long: phosphorescent fish, squid that fly between the waves, twenty-foot-long salps, green medusae. All of them crown a food chain based on the floating microscopic plants. But at the first glimmerings of morning twilight they pause their competition and fade away into the depths. They are gone before the diurnal sea birds even awaken for the day.

Sinking from view, the nocturnal creatures spend the next hours descending through the water column. Finally they settle into their resting schools hundreds of feet down, where there is no light for visual hunters to see them. They adjust to the higher pressure and wait until the sun sets once again. This diel migration is the most massive movement of animals on the face of the Earth. It moves across oceans all over the world, twice each day.

And still another challenge faces the daylight foraging seabirds: back on the nesting island, while they are away at sea, their eggs and chicks are exposed. Feathered opportunists—the frigate birds in the tropics, the skua gulls closer to the poles—stalk the outskirts of nesting colonies, looking to steal unguarded eggs and chicks.

ONE SEABIRD HAS adopted a lifestyle that fits this challenge. Its strategy deals with both the nest predators on land and the retreat of its prey species from the sea surface during the day. That bird is the swallow-tailed gull, a black-headed species with large dark eyes. It appears gray when standing on the rocks, but when it takes flight its forewings and forked tail stand forth in white. It carries a thin circle visible on its back from above—a ring of white tips on dark feathers.

Compared to the routines of most seabirds, this gull has switched shifts. It stays home to brood its egg and defend its nestling by day, and then it heads out to forage across the open ocean in the black of night.

Swallow-tailed gulls nest along rugged coastlines, on slopes above the equatorial Pacific. They prepare a bed of sharp rocks and bleached coral shards, and they carefully lay one egg upon it. This is no down-feathered cup, but it does ensure the egg won't roll away.

The chicks are born knowing how to hide. Their eyes are open and peering out through the first crack in the shell before they hatch. A swallow-tailed gull chick pops from the egg and immediately deserts the exposed area. It stumbles off as it learns to focus its eyes, and it heads away into the rockbound world it is born to. The chick shows itself how to walk in a matter of minutes, and, while its parents remove the shell fragments from the nest platform, it retreats to the shadows.

After dark the parents still know where their chicks are. They find their young late at night when they return to feed them. Should a nest thief drift into view the following morning, the gulls utter a particular call that is picked up and relayed by the other colony members. And should the predator approach closer, a different call sweeps over the colony, and the gulls sprint through the air to mob the interloper and drive it off.

IN THE EVENING, the swallow-tailed gulls gather for their twilight convocation. Their vocalizing escalates till past sunset, then the flock takes wing—not for a twilight spin over the beach like the sandpipers, but for a group flight over a dark and trackless sea.

These birds have oversized eyes, like a Keane painting of a Franklin's gull. Just as the swordfish and the Humboldt squid have outsized eyes to enable foraging at night, so do these gulls have keen night vision.

On the darkest nights, the swallow-tailed gulls are the only birds out there. They have a teeming feeding ground all to themselves. Just as the stars shine through the evening after a hazy, low-visibility afternoon, so does the phosphorescent vitality of the sea come into sharp focus after dark. Like a curtain pulling open, reflections of the sky disappear from the smooth surface of the water, and the depths of the ocean are revealed in perfect clarity. Twenty feet down, phosphorescent outlines of creatures of every stripe appear suspended against the blackness. The gulls find themselves sitting on a transparent medium, looking down at a crowded tableau that begins just below their feet.

The living sea is a world of illumination at night. Some of the smallest creatures in the water are the brightest. They glow like fireflies when they are disturbed, lighting up with blue or green light that lingers even after the disturbance has passed. When herring or squid move through a shoal of these organisms, their movements are illuminated, outlined in glowing plankton. In this light, the gulls can see their preferred prey species.

The bigger fish that may prey upon birds—marlin or sharks—produce a larger phosphorescent disturbance, visible from fifty feet away. The gulls cry out and all take wing when one sees a large predator cutting through the water, leading a glowing wake.

Common lantern fish of all sizes fill the water column. Sitting on the ceiling above them, the gulls are afforded a different perspective from that seen on the wing. From the air, the lantern fish disappear. They are illuminated by light that shines down underneath them, along their ventral side. They control the intensity of the glow they emit, to match that of the faint ambience shining down from above. Looking straight down, a flying tern or booby is unlikely to see this illumination. But it can be seen from the side, by a gull sitting a few feet away on the water.

There is constant motion on the sea surface at night. The gulls watch for something they can snatch as it dashes away from something else. Small fish leap out of the waves to avoid their predators. As they hit the waves again

they turn sharply and speed off. The birds make stabs at them. Sometimes they come up lucky; other times they may catch a fish that tastes of toxic flesh and throw it back.

The gulls are attuned to the rhythms of their prey. The selection of creatures varies on a lunar cycle. The greatest amount of activity below the surface happens on the blackest of nights. The greatest activity above the surface happens with the full moon, when other birds that hunt on the wing—shearwaters, petrels—venture into the dark to join the nocturnal Swallow-tails.

There is moonlight on the water half of the time, usually from a waning or waxing crescent. Moonlight cuts down the transparency of the surface by reflecting back from the waves. It also cuts down on the activity of the small nocturnal creatures, who do not migrate all the way up when the light is greater. The lantern fish are brighter on those nights, but moonlight hunters focus on silver reflections off the sides of the faster-moving fish.

BACK ON SHORE, the swallow-tailed gull chick grows for a few months before it first tests its wings. By the time it flies off its cliff to settle on the waves, the red eye-rings of the breeding adults are fading to black. The parent birds are getting ready to desert the nesting island.

When the chick is strong enough to keep up with them, they all head out to sea. They disappear over the horizon to take up a life on the open ocean. They will remain out of sight of land in every direction for the next half of the year.

NOTES

Most pelagic seabirds forage mostly by daylight (Ballance & Pitman, 1999). Shearwaters and petrels forage at night during the full moon, but the swallow-tailed gull (*Creagrus furcatus*) routinely forages by the dark of the moon (Cruz et al., 2013). Swallow-tailed gulls may stay with their chicks with the full moon, protecting them from predators active on bright nights.

Swallow-tailed gull chicks are precocial, leaving their nest under their own guidance within minutes of hatching. This contrasts with the altricial chicks of passerine birds, whose eyes do not even open for days after they hatch.

The gulls' prey species live near the surface at night, but transition across pressure regimes in their twice-daily vertical migrations. The hydrostatic pressure increases by three atmospheres (forty-five pounds per square inch) for every hundred feet the creatures cross as they move up or down. To manage these changes, the creatures migrate gradually, taking their time in moving through the water column. Creatures that move swiftly across the pressure gradients of depth (diving mammals) use different strategies to cope with the changes of pressure with depth (see "Beaked Whales," Chapter 18).

REFERENCES

Ballance, L., & R. Pitman. 1999. Foraging ecology of tropical seabirds. *Proceedings of the 22nd Ornithological Congress*, Durban, 2057–71.

Cruz, S. M., et al. 2013. At-sea behavior varies with lunar phase in a nocturnal pelagic seabird, the swallow-tailed gull. *PLoS One* 8:e56889.

— 16 —

Young Hawk

THE YOUNG HAWK had finally grown tall enough to see over the edge of her nest. There she found she had been born with an exquisite sense of the world around her. From her treetop eyrie she could see in fine detail for hundreds of miles in every direction. At first she was just looking for her parents to come back and feed her. She could spot them much farther away than she could hear their calls.

As she grew taller, she found she could focus in on vignettes on the forest floor right below her. She could eavesdrop on the smallest elements, as intimately as if she were perched in the branches a few feet above the scene. She found she was particularly fascinated by the smaller ground birds scratching for seeds.

Stippled juvenile plumage emerged from her hatchling down, and the young hawk's wings grew. Finally she stepped off into space to see how those wings could carry her. She found that she was the center of a fluid three-dimensional perspective. And then she learned to take control of that perspective. The green woodland corridors were hers to bank and turn through for miles and miles.

But the hawk was not born knowing how to use her gifts to thrive in these woods. After she left the nest, her parents would feed her when she was nearby—answering her cries for attention. But as she explored farther afield, she spent more time out of their sight. And they spent more of their time out of earshot, following their own pursuits. She would have to learn to fend for herself.

She had left the nest weighing more than her parents, but now she was underweight. She was an agile hunter, but the smaller birds she hunted knew the defensive moves to escape her grasp. Except for the occasional naive fledgling thrush or wren, her prey were not sitting ducks. They worked as hard to stay away from her as she worked to catch them.

THE FOREST AT first light stood deserted, silent and chill, transformed into silhouettes by the mist. Distances across open spaces were foreshortened, the farthest trees faded to obscurity. Droplets beaded the grass and the spiderweb. No birds called, no movement rustled the fallen leaves; there were no shadows; no changing light angles marked the progress of the dawn. When the sun finally appeared it was no more than a small, cold disk against the white—dimming now and then when the fog thickened before it.

A young sparrow rested on a perch where no eye could see her. Any sound, from anywhere around, raised her watchfulness. But the thick wet branches muffled the sounds. She fluffed herself out against the cold and listened through the foliage pressed close on all sides. Cottonwood leaves ticked against each other—a sound like scattered raindrops—when a breeze passed overhead. The sparrow's life could end in the blink of an eye on a day like today, under these trees. The quaking leaves masked the sounds of other motions, including those of stalkers in the bushes.

A meadow waited nearby, faded and yellow. Other sparrows had already found all the grass seed lying under the cover of the undergrowth. The seed that remained lay exposed out there on the ground, beneath open sky. Through the branches, movement caught the sparrow's attention. A few older birds had dropped to the meadow margin to forage. They were hidden by the earth tones of their feathered backs, which matched the texture of straw thatch over wet soil—but they could be spotted when they moved.

These foraging sparrows were furtive, quiet, cautious. With all eyes alert, the chances were better that someone would notice the snake in the grass, the fox in ambush, and call out a warning. With a flurry of wings they flew back to cover, spooked by the flight of one or another of their number who may have seen or imagined something. After nothing further happened they peeped to each other, then reappeared. The boldest going first, they dropped back down to resume their scratching and pecking.

When he had taken the edge off his hunger, the dominant male bird flew into the branches and emerged halfway up the tallest tree. From on high he could see across the meadow and off into the woods. The fog was thinning from the ground up. When he saw that the forest was calm on every side, he began to sing.

The watchman's trills told the other birds that the lookout was clear— that it was safe to move into the open. Soon they had all dropped from the bushes, and three different kinds of sparrows stood shoulder to shoulder. They hopped past each other, looking through the dead grass for their breakfast.

With feet and beaks they pulled aside the fallen leaves, uncovering seeds and occasional slugs or millipedes. Within the protective bubble of song they broke the night's fast, and then they turned their attention to each other. Pecking-order dominance challenges broke out among them as the day brightened.

All around these sparrows the woods came alive. Other kinds of birds, drawn by the spell of the watchman's call, emerged from cover to pursue their own searches. A nuthatch appeared, hunting spiders on a tree trunk, and two kinds of yellow warblers worked the branches farther out. An ornate rufous towhee stepped from the brambles, scratching at the duff on the forest floor.

Another sparrow took a perch up high in the branches, and he too began to sing. After a while, the first one dropped back down to forage, and to reassert his position in the social order. Further back, a thrasher could be heard working the leaves in the undergrowth. A flock of bushtits moved through the branches like a passing breeze. Then, with a flourish of whispering wings, mourning doves touched down in the clearing, while the peal of a flicker sounded across the way, followed by drumming on a hollow root.

The young sparrow had ventured out to forage among her peers and found the food plentiful. But a moment came when her head rose from the inspection of the ground at her feet. She realized that the sentinel song had ended in mid-phrase. No one was perched up in the treetops any longer— all had dropped from sight, leaving only a deepening quiet.

Without thinking, she jumped for cover as did all the other birds around her. A shadow careered behind her just above the ground, slicing through hollows in the undergrowth like a boomerang curving close by. It appeared in an instant, swift and silent but for the rustling of leaves in its wake.

The sparrow felt a collision of sharp points through her feathers, as though she had flown into dried twigs. The impact flipped her against the branches, but she continued her panicked scrambling. She ignored the sting as a few feathers were pulled out, and she pressed deeper into cover, heedless of the thorns and sticks she ran into. Her wings slapped against the bushes until they were too thick to fly through. Then she continued further into the thicket, hopping headlong from perch to perch, until she was buried in deep cover.

The attack passed through the area in the blink of an eye—so close, and moving so fast, that if one bird was captured, none of the others saw. The sparrow wedged herself into a tight space with the foliage pressed close all around. She ignored her bruises. She could see nothing for all the branches, and no eye could see her. She sat and listened for the cheeps of other sparrows, or any sounds of movement through the brush, but the forest stood deserted—silent and still.

A PAIR OF juncos foraged through the glen for seeds and bugs. These birds preferred to hunt in the shadows of the forbs and ferns, but there was not enough food to be found there. The pair had to feed not only themselves but also their chicks, who waited in the shelter of a fallen log. The female had built her nest on the ground there and had incubated the eggs. But as the nestlings grew, both parents were compelled to venture farther and farther into the open, to support the insatiable demands of their young. The chicks cried softly, a call to which their parents were keenly attuned. The cries urged the parent birds to return soon and feed them often.

Now the small dark juncos stood out against the green grass in the clearing. They were torn between paired concerns. How much time should they spend heads-down, inspecting the maze of blades and leaves, versus heads-up, scanning for danger? And how far out could they venture across open ground, given how long it would take them to dash back to shelter?

Juncos are drab birds, colored in shades of the forest floor. Their black eyes cannot be seen against the black feathers of their heads. An observer cannot tell where they are focused—what they have noticed, where they are looking to go next.

When they are on the wing, the white feathers farthest outboard on their tails stand out, more brightly on the males. The juncos steer their zigzag flight using those rectrix feathers; the color shows every time the tail fans or folds. But the central tail feathers—under which the white ones are covered when at rest—are as dark as the body of the bird itself.

The male junco was farthest across the clearing when he sensed movement in his peripheral vision. With his eyes on the opposite sides of his head, this bird has a 360° field of view. He is always aware of his surroundings. What he had seen in the tops of the background bushes was just a blink—a shift in the dappled pattern of bright sky filtering through the branches from behind, as though the leaves were luffing in the breeze.

But there was no breeze. He looked up, focusing on the area, and he saw it again. Another one of the pinpoints of skylight shining through the leaves from behind had winked, as though eclipsed for an instant. The bird twittered to his mate as he jumped into the air. She took off flying only inches from the ground, twittering back, and ducked under a bush.

The young hawk's eyes, both on the front of her face, were focused together, straight ahead, to better judge depth and distance. She had seen the foraging juncos from fifty yards away, and had dropped from her perch to skim the bush-tops and accelerate, all the while concealed behind screens of leaves. Now her targets had flushed, and the hawk was focused on the whiter male; the female had veered off into cover.

The sprinting junco was overmatched by the quickness and agility of the raptor. He knew he would not have time to make it to cover. But he had been pursued like this before and somehow survived. All he could do was what he knew best, so he accelerated at full speed into the top of the nearest bush.

The hawk controlled her hurtling trajectory with the subtlest of adjustments. She guided her final attack to match the erratic flight of her prey move for move. She was flying by reflex—responding faster than thought—locked in on the white tail feathers. They stood out like a beacon against the background, jumping in her vision, shifting around the body they were attached to, marking the target.

When the junco's feet closed on the nearest branch, his wings and tail folded down, and the bird became a running instead of flying creature. His white outer tail feathers slotted beneath the earth tones of the feathers above, and their flashing semaphore disappeared. The appearance of the bird switched from flickering white to the dark drab of a mouse.

In that final half second the hawk's field of vision was a torrent of branches and leaves, light and shadow rushing up to meet her. She was able to keep the glint of her prey's white tail feathers centered in her view. But at the last instant, the white signal vanished. Her scanning eyes reflexively snapped to other passing slivers of brightness in the chaotic background field—highlights reflected from leaves, sun-glints framed by voids in the thicket—all of them dancing ever faster through the passing flow of the foliage.

In milliseconds the hawk realized those other lights were false cues, but before she could recover her real target, she crashed into the branches. The junco had time to take one hop, and he took it sideways. He touched down on a slanting stick—and bounced away in the same motion. He felt the hawk's impact yank the branches under his feet as he pushed off and twisted in midair, diving toward a gap below.

The hawk's outstretched talons absorbed most of the crash; twigs bending against her breast took the rest of the impact. The bush swayed with the collision and rebounded. She was left suspended in branches too thin to support her weight. There was nothing strong enough to perch upon, so she held herself up with wing beats, flying in place while grasping at straws.

She heard the excited twittering of the junco and his mate and looked behind to catch the flash of white tail feathers fleeing into the brush. The small birds called back and forth as they faded into the shadows. The advantage of surprise was gone—the raptor could not pursue them in the thicket. She turned and dropped away, resuming her course through the canyons of green.

THE YOUNG HAWK spent the rest of the morning perched in the highest tree. She listened for the calls of the prey she needed to catch, but the wood was still. She trembled slightly with hunger. She had a lot to master, and only a short time to do it, if she was going to survive.

She would have to learn all about the lives of each of her different prey species while familiarizing herself with the coves and recesses of the mile after square mile of woodlands they lived in. She would have to perfect the nuances of the stalk and the ambush and practice staying focused on a single target when flocks of birds on the wing contracted into tight twisting knots of motion.

She would have to get better at flashing through the narrow spaces without slapping the branches, so close and so fast that the passing contours blur into dappled walls. Then she would have to react instantly when she wheeled around that last sharp turn . . .

Forest-bird-hunting hawks have low fledgling success rates. Most do not survive their first year. Now this one scanned the green landscapes below for signs of life. She was ready to leap from her perch and dive through the trees, accelerating to attack speed. But the longer shadows had flattened into midday—the vista was static. The smaller birds had retired to the shady recesses and disappeared. The forest stood deserted—silent and still.

— 17 —

What Fireweed Knows of Fire

EARTH IS A unique planet. One unique feature of our planet's surface is its glowing vapors. The glow arises from a fourth state of matter—the plasma state. Matter in that state takes the form of a luminous gas. On Earth, plasma is created by combustion. The glowing flame is a plasma— generated by the interaction of free oxygen with hydrocarbons (chains of carbon atoms). Flames are unique to Earth, because free oxygen and hydrocarbons do not occur together on any other (known) planets. Those molecules are produced here by another of the unique features of this planet—the living organisms.

The plasma state is usually hot, but sometimes it's not. Luminous vapors appear on moonless nights in the still air above the bald cypress swamp. Molecules of gas bubble up from the depths and combust spontaneously. Cool blue fire clings to the water. A will-o'-the-wisp builds until the slightest breeze comes to shepherd it across the surface. The radiant apparition slides into the caverns of the trees' buttress roots. Finally it comes to rest, hovering in the darkness. Swelling, then fading, the specter casts the tall boles in silhouette, while the night falls silent.

In the other three states of matter, the chemical bonds that build this world hold firm. The bonds between the pairs of nitrogen atoms in the air, or the hydrogen and oxygen atoms in water molecules, or the silicates in the rocks, are all stable enough to survive the millennia. But in the fourth

[112]

state of matter, those bonds break and release their energy instantly and continually. While the flame glows, the energy that binds carbon molecules together in chains is liberated. Bond energy is converted into cascades of heat and light, and atoms are released from each other.

The fleeting nature of fire is revealed in the candle flame—lucent dark blue below, topped with a rising golden taper. Vaporized hydrocarbon molecules are drawn by the updraft into the blue zone. The molecules are broken down by the heat, and by collisions with oxygen atoms. The bond energy from each molecule is liberated. It drives the chain-reaction destruction of other bonds—producing smaller and smaller fragments. There is enough oxygen at the base of the flame to convert the hydrocarbons all the way to their final, noncombustible CO_2 and H_2O end products.

The atoms in the glowing flame are in the plasma state of matter. They are so reactive that they form and break bonds with each other thousands of times per second. Fantastic compounds flicker into and out of existence, transmuting from one form to another. They cool and polymerize when they rise above the flame. Chains of carbon atoms elongate and branch. Finally they condense into stable lattices of hexagonal rings. The process generates solid black matter made up of a sampling of all the reaction products in the flame. It is the most complex chemical mixture known on Earth—soot.

OURS IS THE only world in the universe known to host organisms that have altered the atmosphere. The presence of oxygen makes Earth the only place where wildfire exists. The waves of heat and light have burned across the landscape ever since green organisms colonized dry land. Plants have long evolved with wildfire, a coexistence marked by the presence of charcoal in the fossil record from hundreds of millions of years ago.

One of those plants is fireweed, a magenta flower that bursts forth all across burned-over hills after the rains. Ordinarily, fireweed is absent among the spring flowers. It is a poor competitor with the rest of the hillside flora for light, or for nutrients in the soil. In order for it to prosper, this plant demands open places where nothing else is growing—a landscape covered with a fresh black layer of mineral nutrients. Those conditions don't exist on most hillsides, where other wildflowers flourish. But full sun and bare ground are present for a few months in winter, every few decades, in the desolate aftermath of fire. In such blackened spaces fireweed grows faster than the rest of the indigenous flora.

To avail themselves of the wide-open, blackened spaces, fireweed seeds resting in the soil have to bide their time until fire has altered the earth above them. But nothing in the silent blackness of their repose changes on the afternoon when wind-driven flames pass overhead. The seeds are buried too deep, after years of interment, to notice the heat or see the light. The superficial conflagration does not immediately alter the chemistry of the living soil in which they rest.

Microorganisms in the soil generate a broad variety of biological molecules. The bouquet of these products imparts the scents we recognize from freshly turned earth. The mix of those hydrocarbon molecules is constantly changing across the seasons. Fireweed seed resting in the ground is bathed in these molecules all the time, and it ignores them all. The seed is primed only to respond to certain small compounds never produced by living organisms. Those unique products are created only in the fire and are available only where the ashes have faded to black.

In a wildfire, the fuel source is cellulose. Its combustion rarely goes to completion. The hot updrafts levitate a blizzard of partially consumed plant fiber fragments up into cooler air. There the breaking of bonds stops, leaving thousands of different molecular fragments of cellulose to solidify from the smoke and drift back to earth. The potpourri of reaction products is different for every fire.

The molecules born in fire can only assemble at temperatures above five hundred degrees Fahrenheit—conditions unknown in living cells. The material cannot be produced biologically; its presence is a sure sign of recent combustion. It is the presence of those particular molecules produced in fire that triggers the germination of fireweed seeds. This plant is attuned to the chemistry of the fourth state of matter.

Fireweed seed can taste the molecules that provide unmistakable evidence of the passage of flames through the grass. In the absence of those specific combustion products, fireweed stays dormant. Much of its seed never detects the particular stimulus to which it is attuned, so it never germinates; it eventually dies. But early in the winter, months after a wildfire, the rainwater percolating down through the soil picks up some of the strange molecules that formed from the luminous state of matter.

The amounts of each of these exotic molecules that settled into the ash are dictated only by random chance. The larger ones may be present only as mere traces. Concentrations of the chemicals that break the dormancy of fireweed seed can be lower than one part per million. But fireweed is exquisitely sensitive to those trigger compounds.

The molecule that brings up fireweed seedlings is called karrikin. It is an unusual fusion of two flat carbon rings, one with five sides, one with six. Karrikin is a trace component of soot—a partial degradation product of cellulose. It is part of the snapshot of the conditions in the flame during that microsecond halfway between the initial ignition of the fuel polymer and the final CO_2 combustion product. When they are bathed in smoke-water carrying dissolved karrikin, fireweed seeds awaken. Later in the winter, they will break through the ground in profusion. A multitude of blossoms will rise three feet high across the hills to replace the black fire scar with a rose-purple carpet.

THE MICROBES IN the soil have also long been adapting to the conditions of their world. They have evolved the capacity to consume any carbon compound they may encounter. They are not daunted by the menagerie of transfigured substances that settle from the smoke of the passing brushfire. They employ their own cold biological fire to metabolize all of these things, growing their numbers in the process. The CO_2 and H_2O end products they generate serve to neutralize the alkalinity of the ash and to humidify the soil in the process.

In the August after the banner fireweed year, fireweed seed showers back down to the ground. But that seed does not encounter the karrikin molecules that prompted it to germinate eight months earlier. By midsummer, the karrikin has disappeared from the soil, consumed by microbial metabolism. In the absence of any stimulus to germinate, the new fireweed seed settles into decades-long dormancy. It rests underground over the years while the soil rebuilds above, and fire-cycle deadwood fuel accumulates above that.

The magenta blanket of fireweed stands pure, free of other colors, during the first spring of the year following the fire. That year is not so good for the other indigenous annual wildflowers. They do not compete for light as aggressively as the fireweed can when growing in fresh ash. The other seeds lie closer to the surface, and some of them would have been killed by the passing flames. The few that manage to grow in the shade of the fireweed set few new seeds.

It will take those other wildflowers years to migrate back into the burn area from seed sources in plants that survived beyond the fire. Other wild-flowers rarely manage the blooming density and vibrancy that the fire summons from fireweed.

NOTES

Three states of matter—solid, liquid, and gas—predominate on the surface of the Earth. The fourth state of matter, plasma, is the most prevalent state everywhere else across the galaxy. The stars are made of plasma, where the electrons fly free and rarely form molecular bonds between atoms. A candle flame is also made of plasma. The molecules in the flame are exotic species (Gaydon & Wolfhard, 1978) found in no other state of matter. During their transient existence, they carry positive or negative charges, or unpaired singlet electrons—conditions that do not exist in solids, liquids, or gasses. A candle flame can be bent by a magnetic field, demonstrating the presence of free electrons within it.

The fourth state of matter is characterized by its glow. The will-o'-the-wisp shines with the spontaneous combustion of swamp gases phosphine and methane. The blue light of that combustion, like that at the base of the candle flame, arises as luminescence from a partial break-down product—carbyne. Carbyne is the smallest molecular fragment of a hydrocarbon. It is made of one carbon and one hydrogen atom. Carbyne is one of the multitude of exotic molecules that only exist in the fourth state of matter. It is exquisitely reactive: once it is created, it lasts only until it collides with another reactive molecule, a matter of picoseconds.

In the core of the candle flame, the combustion does not go to completion. The breakdown reactions at the base of the flame consume all the oxygen, so combustion stops. The partial degradation products condense as they rise, forming particles of carbon black. The particles may reach temperatures of a thousand degrees—they shine in the white light of their own incandescence.

The updraft finally draws these exotic molecules up and out of the candle flame, and they freeze in the form they happen to be in at that moment. As they cool, the bonds between atoms adjust into forms that are stable in the solid state—producing the substance soot. The fire is lit by both luminescence, e.g., from products like carbine, and incandescence. Luminescence is light of specific wavelengths, given off by electrons in orbit around atomic nuclei. Incandescence is blackbody radiation, given off by superheated matter; it appears across the entire color spectrum, its wavelength directly dependent on temperature. Warming matter glows red and brightens to white as it gets hotter. (White is the color generated in our optic nerves in response to a mix of wavelengths of all frequencies.)

Plant fibers are built from cellulose, which is a polymer of glucose molecules. Glucose molecules are built upon hexagonal rings of carbon and oxygen atoms. Each molecule contains twenty-four atoms bonded to one another. Glucose molecules link into a chain to make cellulose. The combustion of cellulose proceeds through the breaking of one bond at a time. At high temperatures, oxygen atoms from the air bind to the carbon atoms in glucose, replacing carbon–carbon bonds with carbon–oxygen bonds. When each carbon atom is finally bonded to two oxygen atoms, and the carbon in each glucose molecule has been converted to six CO_2 molecules, the combustion process is complete. Such combustion has been a unique feature of the surface of this planet for eons (Scott & Glasspool, 2006).

In the fire, the light harvested from the sun by the living plant is given back to the environment, reversing the steps that went into building the complex carbon structures. The atoms are returned to the water and carbon dioxide forms they started from, in a cycle that only happens on Earth.

The figure below depicts the combustion of cellulose on the molecular level. The structure of cellulose is depicted as the polymer of hexagonal rings (glucose molecules) on the diagonal. Such a polymer is degraded by oxygen attack into smaller and smaller combustion fragments, examples of which, including CO_2 and H_2O, are shown. Flame chemistry produces structural rearrangements, which generate a myriad of partial combustion products. These compounds exist for microseconds, then are broken down further to CO_2 and H_2O. One such partial degradation sequence is shown moving left to right on the top line of the figure. A breakdown rearrangement product with fused pentagonal and hexagonal rings is depicted. That product is karrikin. It will be preserved in soot if it passes out of the flame before further degradation. It is produced in several different forms as a byproduct of combustion (Flematti et al., 2011); all of those forms can trigger fireweed seed germination (Nelson et al., 2012). There are many other plants the growth of which is initiated by fire. The magenta carpet of fireweed reflects the complexity of fire ecology.

Another plasma found on Earth as well as on the other planets exists for an instant along the tracks of lightning bolts.

REFERENCES

Flematti, G. R., et al. 2011. Production of the seed germination stimulant karrikinolide from combustion of simple carbohydrates. *Journal of Agricultural and Food Chemistry* 59: 1195–98.

Gaydon, A. G., & H. G. Wolfhard. 1978. *Flames: Their Structure, Radiation, and Temperature*. Chapman & Hall, London.

Nelson, N. C., et al. 2012. Regulation of seed germination and seedling growth by chemical signals from burning vegetation. *Annual Review of Plant Biology* 63: 107–30.

Scott, A. C., & I. A. Glasspool. 2006. The diversification of paleozoic fire systems and fluctuations in atmospheric oxygen concentration. *Proceedings of the National Academy of Sciences* 103, 10861–65.

— 18 —

Beaked Whales

THE SQUID ARE the epitome of molluscan evolution. They have diversified all across the top of the marine food chain. Just beneath the ocean's ceiling, they are big-eyed rockets. They erupt from the waves to fly jet-propelled into the air. Evading predators with their swiftness and agility, they coordinate with each other visually. They signal with colors by day and luminescence by night.

In the deeper sea, the squid don't have the need for speed. They live concealed in the dark and hunt like moving anemones. With arms held out wide, they probe for the prey they will find by touch. They are suction feeders, attacking with tentacles that attach through the negative pressure of their suckers.

We know little about the deep-water squid. Most of their behaviors are hidden from us by the vastness of the sea. They cannot be studied in captivity—they often do not even survive capture. But their success has bred another success story—that of the creatures that prey upon them. We get a hint of the diversity of the squid, and of their impact on the oceanic food web, by our observations of the creatures that have evolved to pursue them—the toothed whales.

The bright-water squid are pursued by equally agile small whales: the dolphins. The larger, mid-water Humboldt squid and giant squid are prey for the largest of the toothed cetaceans, the sperm whales.

Sperm whales have a history as apex predators. They have been the largest, most fearsome creatures the world has ever known, evolving through forms with foot-long sharp teeth—larger than those of the Tyrannosaurs. Those killer sperm whales preyed on the larger fish and the smaller mammals of the high seas. But the ocean did not support life in that niche—theirs was only an evolutionary intermediate from which the sperm whales had to evolve further or go extinct. So over the millennia their jaw slid back along their keel, they lost most of their teeth, their foreheads enlarged, and they evolved into specialist squid hunters. They have now grown even larger in their modern incarnation.

THE OCEANIC PREDATORS that hunt in the twilight depths or in the darkness of night—the larger squid, the swordfish—have larger eyes. The eyes of the cetaceans, however, are proportionately much smaller. Their eyes are positioned on their flanks, not on the front of their heads (like a seal's), where they would provide the best view forward. The bigger toothed whales no longer depend on their eyes the way visual hunters do. They depend on their hearing the way visual hunters depend on their sight.

Their cetacean brains process the echoes around them. Theirs are the largest brains in the animal kingdom. They generate a three-dimensional picture with the details drawn in sound, in real time. They generate the loudest of echolocation calls. They search through the blackness a mile beneath the surface, guided only by the reflected sound waves. They can find their prey miles away.

They generate their sonar internally. The sperm whales produce their pulses in a larynx six feet long. Their wide foreheads are filled with a lens ten feet long, containing an oil that bends sound waves. The sounds they generate can be directed ahead of them like light waves focused through a Fresnel lens.

The sound receivers of the toothed whales are much more extensive than the simple ears of land-dwelling creatures. The sperm whale's slim jaw, more than ten feet long, receives the echoes from its sonar. From that receiver, the information is transmitted up to its inner ear, then on to its brain.

The seabed five thousand feet below the surface is smooth and flat, covered in a layer of silt that has been settling from above for eons. The squid there cannot move behind coral heads for cover like the reef squid far above. They cannot disappear behind a cloud of ink, or by leaping through the surface like the pelagic squid from the photic zone. The defenses the deep-sea squid may deploy against their cetacean predators are unknown.

THE DEEPEST SQUID are preyed upon by the most rarely noticed of the toothed whales. These are the beaked whales. They are not named in the vernacular of the ancient mariners, like the whales more frequently encountered at the surface—the blue, the humpback, the right. Instead, they carry the names of the modern oceanologists who have pursued them out of scientific interest. They include Cuvier's, Blainville's, Perrin's, Stejneger's, Sowerby's, and True's beaked whales. Baird's beaked whale can grow to forty feet long, while the pygmy beaked whale reaches only twelve feet.

There are twenty-two named beaked whale species, with others most likely yet to be discovered. Some of them are known only from a few remains found collapsed on the strandline. We presume their diversity is a reflection of the niche diversity of their prey—the deep-sea squid—but we have little actual data on that.

Like the squid, the beaked whales are suction feeders. Though classed as "toothed whales," they don't have teeth. Similar to the largemouth bass and other salmonoid fishes, these whales capture prey by inhaling them along with the water they swim in. Prey are sucked into the void of a rapidly distending gape and throat. These whales have folds along their neck, similar to the pleats along the throats of the baleen whales, which unfold to aid in that distension.

Beaked whales are the deepest of divers. They stay submerged for hours. They spend a good part of an hour just swimming down to their hunting grounds. They are warm-blooded air-breathers. Their muscles are aerobic at depth, just like they are at the surface.

The deep-water squid they prey upon have a slower metabolism. They are adapted to low oxygen tension and to near-freezing temperatures. The power and endurance of warm-blooded muscles provides the whales with an advantage over these deeper-living prey.

The whales carry with them more than just one deep breath of air, but they cannot carry it as a gas—the volume they would need would be much more than they could carry in their lung spaces. Indeed, the pressure at depth is so great that no air is carried at all—their lungs collapse completely under the pressure through which they hunt.

Instead, like scuba divers, the whales compress the volume of the air they take down with them. They inhale on the surface, save the oxygen, and exhale the nitrogen gas. Excluding the nitrogen decreases the volume of the air they will carry fivefold. They then compress the oxygen by binding the molecules to myoglobin (myoglobin does not bind nitrogen). The bound oxygen takes a solid form, which requires only one thousandth of the volume the oxygen would occupy as a gas.

Myoglobin is an iron protein; the iron colors it red. The muscles of the beaked whales are very deep red, almost black with myoglobin. Beaked whale muscle carries the highest concentrations of myoglobin of any animal on Earth.

The time these whales must spend breathing air on the surface after a long dive is longer, the longer they were down. When they stay down at depth for hours, they have to rest on the surface afterward for an equivalent time, gradually reoxygenating their myoglobin. After breathing once per minute for an hour, they will have accumulated hundreds of deep breaths worth of air to carry down with them when they sound.

WHEN THE BEAKED whales go hunting, they point their heads toward a bottom hidden by gray darkness. As the world of light recedes, they fold their pectoral fins into dents on their bodies (called flipper pockets) and pump with their tails, accelerating straight down. Brightness fades from their peripheral vision; they give up the light during the first minute of their descent. Then they dive vertically for half an hour or more, headed for their benthic haunts.

Blinded by the darkness, the whales lose their visual sense of "down," and their path wanders away from the vertical. The world ahead goes black enough that they come to notice phosphenes—random background flashes that arise as visual noise in their own optic nerves. The pressure upon them increases steadily. Concentric circles of color may momentarily blossom into view on their retinas when the motion of their heads create pressure waves in the liquid inside their eyeballs.

Eventually some of the fainter spots of light they perceive appear to float across their visual field. Sometimes those lights jump to the left or right. Some of them flash on and off. These are the lights from the realm of the luminescent creatures that live in the darkness of the midwater depths. Sinuous strands of stars move against the background in train, trailing from glowing siphonophores or salps. Little fireflies bob and jiggle—the lures of giant oarfish. Schools of lantern fish, invisible from above, become a lighted ceiling suspended in the water—after the whales dive through their ranks and can see their glowing undersides. As the dive continues, the lantern fish shrink away into the dim ambient light that lingers at the zenith above.

These whales enter a hunting ground they have all to themselves. None of their own predators hunt there. The whales click several times per second as they move. They issue trains of sonar clicks that increase in frequency. Those clicks bounce back as echoes smeared out into ascending or descending tones, modulated by reflection from the diffuse thermocline layers in the water. Each whale also hears echoes from the calls of the other members of its diving pod. The returns are softened and distorted by distance. All the echoes add to the mental picture these animals create of their habitat.

THE BEAKED WHALES dive into the aphotic zone, the domain of utter darkness. As they descend, they disappear from our sight, and from our understanding.

We are creatures of the light. We understand our world through what the light illuminates for us. The deep-diving whales leave that world behind as they plunge away into the depths. They move into a realm in which eyes offer no guidance. Our own understanding of their lives is limited by our inability to comprehend their world as they see it.

These whales illuminate (ensonify) their ocean with sound. Sound travels through water hundreds of times farther than does light. Whale echolocation calls can be detected miles away from their source. The whales may well see their world in much greater depth than we can even imagine. Their overview of the sea floor may correspond to our view of the mountains from a vantage point flying thousands of feet above the ground.

We do not know how much detail echo-visualization can show across such great distances. The speed of sound in water is four times as fast as in air. The density of the water is greater at depth; the pressure increases to tons per square inch. This sharpens the details carried by sound. Sound waves deform softer surfaces, which affects the way the waves are reflected from them. The whales may be able to sense the textures in their world, including the textures of the creatures they hunt, to a degree that exceeds anything in our own experience.

THE SHADOW ZONE, a mile below the surface, is a place in the ocean where the oxygen levels have fallen too low to for most oxygen-breathing creatures to survive. But the vampire squid are adapted to life in this low oxygen environment. We do not know how numerous they are across these areas. The deep diving whales, however, dive to these depths. They bring their oxygen with them, giving them the capacity to hunt through this anaerobic zone.

In the lore of the marine biologists, Count Dracula is the namesake of the vampire squid. The Count is known for his ability to draw his cape around him, change his form, and disappear into the night. The vampire squid has that same capacity. It can draw its eight webbed arms around itself and disappear under the webbing. We know what this looks like in visible light: the squid appears to transform into a sphere—the contours of which are obscured by circumferential rows of soft, flexible fingers. However, we do not know what this looks like when illuminated by sound. Does the squid disappear? Is this behavior an adaptation that allows it to escape cetacean predation? Are some of the whales specialists who can deploy a predatory counterstrategy that involves sound?

We have much to learn about the beaked whales and their prey, but they are among the most difficult creatures on Earth to study. They are concealed by great distances that intervene between them and those who study them. They inhabit a landscape about which we know less than we know about the surface of the moon. Whale vocalizations can be monitored by hydrophones. They can be tagged, and their dives monitored by sonar. But the details of their lives are hidden behind marine layers that distort the sounds we listen for. For now, they are mysteries that ride beyond the limits of our knowledge of our world.

NOTES

We face a particular challenge in understanding behaviors of animals who perceive the world through primary senses different from ours. Examples of perceptions we cannot describe based on our own experience include the olfactory world of the canines, vibrational signaling in plant hoppers, pressure sensory fishes, and acoustic visualization in bats, some birds, and whales. When we cannot see the forest for the darkness, we illuminate the trees with a light—staying with our primary visual sense. The whales ensonify the darkness, process all the reflected echoes, and form a mental picture of their surroundings. It is a challenge for us to see how that would be done. The brains of the toothed whales, which process the acoustic signals, are proportionately the largest brains in the animal kingdom, larger than those of the baleen whales. The whales' hearing is comprehensive—they hear ultrasonics: sounds up to 30 kHz (humans can't hear anything over 20 kHz). The beaked whales routinely produce five clicks per second (Johnson et al., 2004). They increase the rate of their sonic pulses when solving the two-body problem of their and their prey's motions as they close the intervening distance (Au et al., 2013); they filter out the louder outgoing pulses and listen for the fainter returning echoes. This is evolution parallel to the same solution used for the same problem by the bats.

The melon is an organ on the heads of the toothed whales that focuses their sound emissions forward. It is filled with liquid spermaceti—a term derived from two words: the family (cetus, in its possessive form) and species name of the sperm whale. The characteristics of spermaceti are such that the melon may solidify in contact with the near-freezing temperatures of the deep sea. This would theoretically counter the buoyancy of the animal's head as it descended into denser water. But what effect would that have on its use in signal generation?

The beaked whales are classed as Odontoceti, together with the rest of the toothed whales, and separate from the Mysticeti (baleen whales). But beaked whales don't have teeth, with the exception of adult males, which have a pair of tusks that point up from the lower jaw and appear to be sexual display (sometimes they are adorned with barnacles). Beaked whale species are still being discovered. Their behaviors are as yet poorly documented. A Cuvier's beaked whale was observed diving to 9,300 feet, during a dive that lasted two and a quarter hours (Schorr et al., 2014).

Beaked whales are extreme divers (Tyack et al., 2006). They express extreme levels of the oxygen-binding protein myoglobin in their muscles (Helbo & Fago, 2012) which enhances their deep diving capacity.

Vampire squid (*Vampyroteuthis*) live in anoxic regions of the deep. They forage off of marine snow, which is biological detritus that settles from above. Their defenses include an ejection of luminous clouds of material and a retreat into a spherical shape wrapped up in the webbing between their eight arms. Their reaction to attack by beaked whales is at this time unknown.

The bends attacks scuba divers when gas exsolves in their veins as they ascend to lower pressures while breathing pressurized air. Whales do not get the bends because they do not breathe at depth. They dive on a single breath of air, which compresses with them when they dive and decompresses when they rise.

Divers who have been ensonified by sperm whales have been impressed by the sensation of the sonic waves passing through their chests. This may have led to the notion that sperm whales could focus great pulses of sonic pressure on their prey with enough energy to physically stun them. But there is no documentation to support that.

Prehistoric sperm whales had teeth on both upper and lower jaws, which were advanced to the tip of the snout. They were predators of large marine fish and mammals (Bainucci & Landini, 2006). One specimen the size of the modern sperm whale has been inferred from the recovery of a single thirty-six-centimeter-long tooth (Lambert et al., 2010).

The zone of living phosphorescence, where the creatures generate their own light, begins just a few hundred feet down, where the ambient light fades away. This living phosphorescence rises to the surface level at night. Long siphonophores are luminescent at depth (Haddock et al., 2005). Some lights in the dark arise as phosphenes in the eyes of the beholder. Some of those are the result of pressure waves within eyes, generated by a flick of the head (Nebel, 1957).

REFERENCES

Au, W. W. L., et al. 2013. Nighttime foraging in deep diving echolocating Odontocetes off the Hawaiian islands of Kauai and Ni'ihau as determined by passive acoustic monitors. *Journal of the Acoustical Society of America* 133, 3119–27.

Bianucci, G., & W. Landini. 2006. Killer sperm whale: A new basil physeteroid from the late miocene of Italy. *Zoological Journal of the Linnean Society* 148: 103–31.

Haddock, S. H. D., et al. 2005. Bioluminescent and red-fluorescent lures in a deep sea siphonophore. *Science* 309, 263.

Helbo, S., & A. Fago. 2012. Functional properties of myoglobin from five whale species with different diving capabilities. *Journal of Experimental Biology* 215, 3403–10.

Johnson, M., et al. 2004. Beaked whales echolocate on prey. *Proceedings of the Royal Society B* 271, supp 6, S383–86.

Lambat, S., et al. 2010. The giant bite of a new raptorial sperm whale from the Miocene epoch of Peru. *Nature* 466, 105–8.

Nebel, B. R. 1957. The phosphene of quick eye motion. *Archives of Ophthalmology* 58, 235–43.

Schorr, G. S., et al. 2014. First long-term behavioral records from Cuvier's beaked whales (*Ziphius cavirostris*). *PLoS ONE* 9: e92633.

Tyack, P. L., et al. 2006. Extreme diving of beaked whales. *Journal of Experimental Biology* 209, 4238–53.

— 19 —

Raising the Bridge

THE LAND ANIMALS were gradually growing aware of a distant ridgeline standing over the sea. When they lifted their gaze across the ocean, if the visibility was particularly good, a low rolling silhouette would sometimes materialize on the horizon. At first it was a subliminal presence—easily forgotten after a rare appearance on a clear day. But as the generations passed, the creatures came to accept those contours as permanent features above the oceanic horizon.

The ground sloths and glyptodonts paused in their wanderings along the northern reaches of the southern lands. They could see a peak over the water, rising against the sky. Streamers of cloud often trailed from the distant crest—white condensation from the trade winds—or darker billows of ash.

A very different community of animals lived beneath that peak, on the southern edge of the northern lands. From high up on its flanks, the peccaries, gomphotheres, and coatimundis sometimes saw similar peaks rising above the ocean to the south. Out across the waters a new land beckoned.

At times when the browse was thin, or after a territorial fight, some of these creatures would raise their sights out across the sea. They were drawn by the promise of ungrazed fields to forage, and unclaimed space to occupy and defend. But oceanic currents prevented even the strongest swimmers— the otters, the alligators—from crossing the straits between the volcanic peaks. All those that set off from the shore eventually returned—exhausted, miles to the east of where they had started.

And yet some plants had already made the crossing. The drift-seeded shoreline species—mangrove, beach almond, nickernut—were already common above the tidelines of both opposing shores. Many birds and bats also knew both sides of this strait. Some had come to divide their time between the two lands, commuting across in seasonal migrations.

Day-flying moths and night-flying butterflies navigated the air lanes over the channel to discover unexploited niches free of their predators. The numbers of those lepidopterans would then explode above the new habitat. Their periodic mass migrations darkened the skies above the water for days.

Some of the traveling birds carried plant seeds in their guts, left over from their last meal before the crossing flight. When they later excreted them, those seeds took root where they landed. Initially, upland plants grew from seeds carried over the water by the smaller migratory birds. Later, as the water gap closed, larger birds with shorter migrations—macaws and toucans—transported the larger seeds from the lowland jungles.

As the birds continued their explorations, they expanded the ranges of plants coming down from the north and up from the south. Both sides of the water gap came to support similar forests, growing similar trees. Those forests shared the same weather and soil, but very different land animals walked in their shade—animals from different continents.

THE TWO VOLCANIC peaks stood above their respective shores, facing each other across the strait that separated their two lands. A straight line ran across the water between the two monoliths and continued beyond them in both directions to connect a string of other conical peaks. The volcanic mountain range extended over the horizon to both north and south. The line of rising peaks paralleled another line that lay under the water fifty miles to the west, out in the Pacific Ocean.

That offshore line tracked along an offshore trench in the sea bottom. The sea floor west of the trench was moving east at a rate of one centimeter per year. As it advanced, this moving ground met a stationary block of the Earth's crust. Where the two plates of the crust met, the block from the west slid into the trench and dove down to continue its easterly advance underneath the stationary block.

As it slanted downward, this subducted block grew warmer. Fifty miles east of the offshore trench, it had descended twenty miles below the surface, and its rocks melted from the heat and pressure.

The molten rock was buoyant—less dense than the rocks around it. Over tens of thousands of years it accumulated into incandescent bubbles miles in diameter. They pushed fissures through the rocks as they rose upward. When they neared the surface, the rupture of each successive subterranean fissure sent earthquakes rumbling upward and out across the ground.

When the upper edges of these bubbles broke through the sea floor, their tremendous heat boiled the water. The ocean steamed for miles in every direction. Liquid rock spilled through fissures and piled up to build peaks that would emerge above the waves. A north-south arc of volcanic islands arose from the sea along a line fifty miles east of the offshore subduction trench.

OVER THE MILLENNIA, the animals from successive generations had watched the offshore volcanoes emerging in line, one after the other—each closer than the last. The sea smoked, and new peaks arose in front of the older peaks, and reached toward the stratosphere. The glow of the Earth's red blood streamed down their flanks through the night, at the end of a ruddy path of reflection across the water.

The rising peaks intercepted the tropical flow of the trade winds, precipitating deluges on their windward slopes. Cataract canyons cut through basalt ridges, carrying sediments down and across black-sand shorelines. Shallow sedimentary shelves built up beyond the beaches.

The Caribbean Sea lay to the east of this arc of volcanic islands. The western edge of the Caribbean plate rose in response to the tectonic movement beneath it, uplifting those near-shore shelves until they broke the sea surface. The uplifted land bridged the seaways between the volcanoes. The plants advanced to cover the new saddles linking the young islands. Then the animals advanced under their shade, along the newly risen path across the water.

The Pliocene creatures of North and South America slowly found their ways closer to one another. Camels, raccoons, coyotes, wolves, bears, and horses extended their ranges to the south as the sequence of rising volcanic islands linked the newer to the older. And from the other side, a unique ground fauna was advancing to the north.

Up until that bridging moment in geologic time, the continent of South America had been an island. It carried its own mammalian fauna, derived from the ancient continent of Gondwanaland. These were marsupials. They filled the niches in South America that placental mammals of that time occupied to the north. The northern fauna were derived from a different ancient continent—Laurasia.

The marsupial South American fauna included carnivores—saber-toothed cats, shrew opossums, an aquatic wolverine. They preyed upon large herbivores such as the ground sloth and mammals that had evolved to fill the niches of the elephant and rhinoceros. Descendants of these creatures waited on the north shore of South America for the last strait to close and finally link their lands with the continent to the north.

AS THE CENTRAL American land bridge rose, it changed the ocean currents around it. Many millions of years earlier, the southern tip of South America had lost a similar land bridge, connecting to a continent further south—its last connection to Gondwanaland. That bridge to Antarctica dissolved into isolated islands and fell back into the sea. The currents swelled through the opening straits and bent their course into a ring around that southernmost land. The ring current cut off access to warmer waters, condemning the southernmost continent to a future of permanent snow and ice.

Now the currents above the northern tip of South America would also be changing. The oceanic flow between the Pacific and the Atlantic would be cut off by the rising land bridge. The warm Atlantic equatorial current would be deflected further north, to follow along North America's eastern seaboard. This warmed the North Atlantic, bringing more rain to Greenland and Northern Europe, and more snow farther north, in the Arctic.

These changes in the oceanic currents would alter the climate. More snow at the two poles would reflect away more sunlight, lowering the global temperature. After the isthmus connected the Americas and separated the oceans, a sequence of ice ages would begin.

FINALLY A MIASMA of smoke and sparks exploded in the middle of the channel separating North and South America. Thunderous rumblings carried across the water for weeks. Smoke and steam obscured the far shores. When the quiet returned, the initial dome of a new cinder cone was revealed above the waves in the center of the strait. Its eruptions would alternate with years of quietude, but successive outbursts would eventually bring dark conical symmetry to this rising island. It would fall into line with its sister peaks to the north and south.

Coconut palms, screw pines, and sea beans sprang up to fringe the new volcano, giving the migratory birds a resting place on their commute between the continents. Over the decades, low growth took hold behind the beach, then climbed the slope into upland forest. Stripes of this green were occasionally paved over in gray by volcanic outbursts. Avalanches of fiery ash accelerated down steep flanks and slid into the water, broadening the base of the mountain toward the shores north and south. But the plants grew across these dead zones faster than the eruptions covered them, and the rising peak remained green.

The waters to the north and south of the volcano grew shallow, and the current slackened. The larger animals were able to swim to the volcanic island. They explored the shore and the forest, and the transient crater lake at its peak. They found other creatures, strange to them, resting in the shade. But they had always been most interested in the lands beyond the far shore. So they moved on, swimming the rest of the channel and hauling their dripping bodies out to explore new habitats.

THE SUBDUCTION OF the Pacific Plate beneath the Caribbean Plate continues, and the volcanoes above the subducted margin remain active to this day. They grow two miles high, until they sink into the crust under their own weight. The molten conduits at their cores branch toward the lower resistance of side channels on their flanks. New cones appear in the foothills and in the saddles between older peaks that have grown dormant.

A grand basaltic cordillera now anchors the spine of Central America. Rivers filled by the drenching rain of hurricanes flood down its slopes. They transport sediments that broaden rainforest plains to the east and west. Animals and plants from north and south have mixed on the wide land bridge. They occupy one of the most diverse habitats on Earth, at the nexus between two biologically unique continents.

To the north, the large South American grazers had moved into the great plains. And species from the north—weasels, foxes, jaguars, puma—had arrived in the south to displace the formerly dominant marsupial predators.

This great interchange was the most extreme biogeological upheaval to be seen in recent history—up until the event that followed it.

Today only the smaller migrants from South America—the armadillo, porcupine, and possum—survive in the north. A few of the migrants from the north have survived in the equatorial jungles of their newer home: capybara, tapir, the large cats, the small canids. But all the largest of the New World herbivore fauna to cross the bridge have disappeared, along with the Pleistocene predators that followed them.

This wave of extinctions happened at the dawn of the Anthropocene. The great loss of animal diversity came at the hands of another migrant that entered North America across another land bridge, moving east from Asia. This species, the Paleoindians, destroyed the novel animals that had come up from South America. Then they migrated across the Central American land bridge and swept away all of the larger animals they encountered in the neotropics, all the way to the southern tip of the southern continent.

NOTES

There is also biogeologic evidence for another land bridge that rose prior to the rise of Central America. That first bridge arose from the subduction zone on the eastern edge of the Caribbean Plate. Its high points remain above the sea in the form of the Antilles islands. So, two bridges hosted the two-way migration of species between the Americas. This great interchange of terrestrial animals was first perceived by Alfred Russell Wallace (Simpson, 1980).

The current land bridge still retains pockets of very diverse fauna. There are more species of birds around some of the volcanic mountains in Central America than there are in all of North America. The insect diversity includes the night-flying butterflies of the genus *Caligo*, the owl butterflies. Massive insect migrations were known as recently as the time of Columbus, by day-flying moths of the genus *Urania*, the sunset moths.

Simpson, G. G. 1980. *Splendid Isolation: The Curious History of South American Mammals.* Yale University Press: New Haven.

— 20 —

The Color of Sand

A BAREFOOT WALK down the beach can take you back to the past. The sand between your toes describes a history of the uplands above your shoreline. It is a record that goes back millions of years, with the color of the sand as the first part of the story. For instance, the grains in a salt-white beach are tiny glassine marbles. They are bits of quartz that was forged long ago, miles below ground level. That quartz was born from a bubble of incandescent magma miles in diameter. The bubble eventually became part of a continental mountain range.

It took millennia for the molten bubble to cool as it rose from deep in the Earth's crust. The subterranean monolith finally solidified into granite, under tremendous pressure, still miles below ground level. That pluton would not see the light of day until eons of erosion removed the softer rocks above it. Then the exposed granite itself would begin to erode—much more slowly than did the sedimentary strata from which it emerged.

Chips of granite domes spall away over the years, along with pieces of every other kind of rock that makes up the mountains. They fall into alpine streams to begin their rough and tumble journey down the rivers to the sea. In the water, most of these fragments are reduced to dust (silt). But white quartz in granite—formed under tremendous pressure—is one of the hardest of minerals. It does not dissolve into stream water. It survives the rapids.

As their sharp edges are worn back, the pieces of quartz grow smaller and more rounded. In that form they pass through the hundreds of miles of flash floods and cascades that lead them through the canyons and deltas to the sea. Some of these tiny crystal balls in the beach sand have made this journey many times.

The smaller they grow, the more resilient these quartz grains become. When they have shrunken down to a millimeter or less in diameter, they have grown spherical. They have lost the weight that would grind their surfaces on impact, and also lost the ridges that would catch on whatever they might roll against. At that size these particles have reached their evolutionary climax: they will last like that unchanged while they move through geologic time across successive generations of mountain streams.

Some of the grains in the beach at your feet are a billion years old. They roll up and down the shore for decades before they are pulled farther out to settle on the seafloor. There they are buried and compacted into stone. Eventually, their strata are exhumed and uplifted into ridges that rise above the water to join the continental margins.

Once those rocks are raised into the air, they are subject to erosion. Quartz grains find themselves in the walls of young canyons. Steep cliffs focus the wind or guide the waterfalls that liberate the grains from their sedimentary matrix. Then they are freed to begin their journey toward the sea once again. By now, they have flowed down many rivers, over thousands of miles of landscapes, most of which have been eroded down into oblivion and no longer exist.

Sliding silicate blankets of sand have accumulated in coves and dunes over geologic time to become a permanent part of the Earth's surface. Quartz sand in the rivers or on the beach represents a minute fraction of all the sand locked up in sandstone, waiting to be released by wind or water for another brief moment under the sun.

The sand is sometimes accompanied in its migrations by other minerals. It sparkles with gold in the river where it shares the channel with square pyrite flecks (which will not survive even one cycle of passage to the sea). It is edged in amethyst on the shore beneath uplifted rocks that release garnet as they weather away, such as on Pfeiffer Beach in Big Sur. There, on occasion, purple rafts of beached by-the-wind sailors follow the strandline, adding to the purple braids on the sand beneath them.

ISLANDS REMOVED FROM continental shores by thousands of miles of ocean, far from the snowy granite spines of the mountains, still support crescents of beach beneath palm-shaded shorelines. Silicate spheres do not float across the sea; these are not the same as the beaches that border the mainland. Under bare feet, tropical sands are softer, almost powdery. They are made of grains that continually grow smaller, since they are slightly soluble in seawater. They shrink to the size of dust particles, and finally disappear, in a matter of mere decades.

Just as quickly, they are replaced. Such tropical beaches are made of biogenic sand—derived from coral. Corals grow by recycling dissolved carbon dioxide into the solid calcium carbonate gardens of stone that decorate the reef. Their photosynthetic symbionts thrive where intense tropical sunlight penetrates the water.

Coral provides a solid foundation for the reef and a habitat for the reef's many residents. Some of these inhabitants destroy coral. A principal culprit is the parrotfish. These beasts break off chunks of the living coral and crush it with the grinders in their throats. They extract the nutrients from the coral polyps, then at the end of the digestive process, they excrete ground carbonate fragments as sand. The biggest of these grazers produce hundreds of pounds of coral sand per year, in a process that has been continuing across geologic time.

Those bits of ground coral stonework accumulate on the bottom in drifts large enough to be braided by the tides. Storm surf sweeps the coral sand up above the waterline. There the waves round off the sharper edges as they begin the process of reducing the grains back into soluble carbonate.

In contrast to silicate sand, which travels hundreds of miles downstream to reach a continental shoreline, carbonate sand travels only a few hundred feet to reach its crescent beach. Its depth along the strand is a measure of the productivity of a tropical reef. Where it slides away to settle into deeper waters, it will eventually compact into limestone.

This biogenic mid-oceanic sand forms within 30° of the equator. It contains a mélange of carbonate forms. Aside from bits of coral and flakes of coralline algae, these include broken mollusk shells, urchin spines, sea sponge spicules, and calcareous diatom tests. The microscopic asteroids in the mix are the hard parts of foraminifera, which build their solid disks where they live on the seabed.

The tropical beach does not sparkle with the reflected crystalline glints of silicate sand. But coral sand does reflect its own source materials. Caribbean and Polynesian beaches are often shaded in the pastels or saffrons of the particular corals that grow offshore. The pink sands of Bermuda are tinted with the tests of a particular foraminiferan that thrives in the shallows of the western Atlantic. Each of the biogenic beaches from the Seychelles to Trinidad wears its own particular off-white shade, a reflection of the living reef from which it arose.

OTHER MID-OCEAN ISLANDS are not coral and limestone, but volcanic. These include shield volcanoes, which stand where magma from Earth's mantle flows up through rents in the crust. Layer upon layer of basalt piles up on the sea floor, twenty thousand feet deep or more, creating masses of land the size of mainland counties. These mountains include Ascension Island, the Comoros, and Hawaii.

At that height, these dark ridges make their own rain. Where the trade winds ride up the slope, humidity on the windward side is borne into the stratosphere. Rain clouds condense at the crest, and the runoff carves furrows into the volcanic flanks. Smooth lava flows pleat into successive valleys, an archetype of which is Waimea Canyon.

The magma derived from the mantle below these islands forms colored layers in its successive strata. The reds, oranges, and blacks of those layers are revealed in cross section on the walls of Haleakala Crater, on Maui. Those layers color the sand that is eroded from higher-elevation cliffs and carried down-canyon with the runoff from the rain.

Black-sand beaches on the south shore of Hawaii are made of basalt grains stained with iron from the mantle. Where the iron has been oxidized, the sand takes on the brick-red color of cinder cones. Mantle-derived olivine releases green sand to fill some coves hidden beneath the island's cliffs.

Basalt is softer and more easily eroded than metamorphic continental rocks. Kauai rose to its highest, almost thirty thousand feet above the seabed, some five million years ago. Since the volcanic vents have gone dormant, the peak has been lowered to a quarter of its maximum height above the waterline. Most of the broad shield volcano has already been reduced to sand.

Kauai's shores show a mix of carbonate sand, generated from coral reefs offshore, and basaltic sand derived from the island's volcanic spine. The beaches there can vary from auburn to blond in color, where the darker red and yellow iron-stained sand is lightened by white coral bits blended in by the waves. The shades of the sand where the surf runs up the slope will change as you walk barefoot from one beach to the next, on a trip back through geologic time along the Kauaian shoreline.

NOTES

Quartz is the hardest mineral commonly encountered on the surface of the planet. The strongest geometrical shape a grain of mineral can assume is that of a sphere. As quartz grains are reduced in size by geological processes of tumbling against each other, their shape approaches the spherical. Their strength-to-mass ratio increases as they shrink, until they reach the size of sand grains. At that stage, quartz sand resists any further reduction in its size. It has become immortalized: it will continue to cycle, forever unaltered, into and out of sandstone, across geologic time.

Bumphead Parrotfish grow to more than a meter in length. Adults have been calculated to produce hundreds of pounds of carbonate sand per year—each—as a byproduct of grazing on coral (Thurman & Webber, 1991). Tropical carbonate sand also contains microscopic asteroids: star-shaped carbonate disks called tests, produced by the foraminifera.

I thank Siim Sepp, creator of Sandatlas.org, for his input to this story.

REFERENCES

Thurman, H., & H. Webber. 1991. *Marine Biology* (chapter 12) HarperCollins, NY.

— 21 —

Dust to Dust

MOTES OF DUST drift weightless in a shaft of sunlight slanting through the window. There in the morning calm our mind can drift as well—out the window, up into space, and back in time. We can wander for billions of years, back to the birth of the solar system—to where there were no grand planets to support creatures in living rooms, only drifting motes of weightless dust.

Way back then, the first light of the newborn sun exploded with a storm of particles. Electrons and protons flew away at millions of miles per hour. At that speed they encountered the chill of deep space beyond the orbit of Mars in a matter of days. Then they came together and bonded to produce atoms of hydrogen. Carbon and oxygen atoms also formed in this solar wind. Finally, oxygen and hydrogen atoms bonded to form water. Carbon atoms chained themselves together into ever-longer strings.

Those thin strands of carbon grew as they flew, lengthening into filaments of dust. As they receded from the heat of the sun they passed the frost line (now beyond the orbit of Jupiter). At that distance water molecules froze, and the newborn strands of dust grew a coating of ice.

As it receded into space, this frozen dust became mixed with the cosmic dust that pervades the galaxy. The pressure of the sun's light pushed it all away, but the pull of the sun's gravity drew it back. The dust offered no resistance—it collected where the forces upon it balanced.

Over time, some of this dust consolidated into clumps, which condensed into balls. They enlarged over the eons—their growing masses amassed enough gravity to attract ever more stardust. They swept the dust from their orbital tracks through space onto their surfaces. The weights of these spheres grew to crush the filaments at their cores into solid matter. Their surfaces above took on planetary contours.

When they had grown large enough, ice melted into water on some of these newly forming protoplanets. Barren flats on solid surfaces disappeared under shallow ponds. In the wet depressions, chemistry driven by sunlight linked the polymers of carbon atoms into wonderfully complex compounds. Those various ponds became niches, in which simple life forms would arise and then keep on evolving, toward the living diversity of animals and plants we see around us today.

THE COSMOS IS a dusty place. Clouds of dust have been accumulating in the galaxy since its creation. Stardust solidifies in its simplest of forms, and then it remains that way for billions of years. Today's floating space dust is largely unchanged from the forms it took when it first appeared, back at the beginning of time.

Stardust enough to make millions of planets has accumulated since then. It is pushed around by starlight into the recesses of the galaxy. We can see this dust when our eyes adjust to the night sky. It appears as dark lanes that course in front of the star clouds of the Milky Way.

Space dust forms from carbon atoms heated to hundreds of thousands of degrees. Single atoms of carbon like those do not exist on Earth. Such singlets are very reactive, and should one appear here, it would combine with the nearest air molecule in femtoseconds.

Where they meet, as they cool in the vacuum of space, single carbon atoms combine. Each atom binds to three others of its kind, eventually forming hexagonal molecules that condense into graphite. If there are free silicon atoms in the surrounding space, other carbon atoms bind to them, forming crystals of silicon carbide.

Earth formed from space dust. That dust has since been processed on Earth during its long residence here. The outdoor dust on Earth—the material that billows backlit behind a jeep driven down a dusty road—has had its carbon removed. It is silicate that has been weathered into clay, then dried and powdered—only the heavier elements of space dust remain: silicon, oxygen, aluminum, magnesium.

THE MOTES OF dust floating in the sunlight under the window respond to the same forces as do the particles of cosmic dust. They collide and stick together, consolidating into clumps that settle, under gravity, to the floor. Then they swirl away from passing air currents into dark, out-of-the-way recesses.

The dust in the house is also stardust. It is mostly carbon, built of atoms that were blown from the surfaces of stars billions of years ago. But house dust is a highly modified form of its cosmic precursor.

The thin layer of dust on the inside of the window glass may look just like the layer of dust on the outer side. But the indoor dust has arisen from a process vastly more involved than the one that produced the dust outside. House dust is incredibly more complex and organized than the dust on the road, or the weightless dry matter drifting between the stars. Star dust is one of the most common constituents of galaxy, whereas house dust is the rarest substance in the universe—occurring in trace amounts on only one single planet.

House dust is formed by biological reactions. It has been processed to enrich its carbon. It contains polymers of amino acids, the protein sequences of which are specified by genetic information. That information was acquired and refined over billions of years of evolution and has been encoded in DNA. That DNA is also contained in house dust.

House dust is made of the microscopic dead cells that dry and then spall from our arms and legs under the passage of sleeves and cuffs, finally to fall off into space. Raise your hands into the shaft of sun shining through the window, run them along your forearms, and watch the backlit airborne dust particles multiply and then sink down toward the floor.

House dust is part of the living world. It has the capacity to support life. Its structure and composition are always in flux. It aggregates to build one of the world's many niches, which is soon occupied by one of the world's many living communities.

Inside dusts arose only after living organisms arose—and after they constructed their own living spaces. It supports an ecology of creatures that recycle indoor dust back through the outside world. In contrast, the graphite flakes of space dust that fall into the atmosphere, and the silicates of outdoor dust, are inert. They are not metabolizable by any life forms, not part of the cycles of the biosphere.

THE LIVING WORLD on Earth has evolved to take advantage of the opportunities that arise to it. Like the myriad of other niches on Earth, the dust bunnies scurrying across the floor support their own self-contained microcosms. Animals now live everywhere—in the woven downy baskets of deserted bird's nests, or around the furred edges of mouse holes underground. When human habitations appeared, the animals from those ecologies moved indoors to occupy the dust niche, to which they were already well adapted. A diverse collection of creatures came prepared to occupy the desert beneath the dresser—one of Earth's most barren environments, but still habitable by some.

The indoor dust niche is inhabited by a variety of highly adapted arthropods. Any water vapor that may absorb to the amorphous tangle of microscopic filaments evaporates away again immediately. Water is essential for all life on this planet—including the species inhabiting the fuzz under the refrigerator. But mist enough to drink would dampen this dust into a mud that would smother its inhabitants.

Instead of depending on liquid water, the smallest dust creatures acquire water in its gaseous form, vapor, from the air. Or they synthesize the water they need internally, feeding on dry matter and generating water as a byproduct of their own metabolism.

A pyramid of life in the house dust is based on microscopic mites and fungi, which can directly consume the edible matrix. A million dust mites can occupy the living-room carpet. They share the space with larval forms of carpet beetles, drugstore beetles, dermestid beetles. They are small, pin-cushion forms with no visible appendages; they feed on the dried filaments. Some lepidopterans are also part of this guild. Clothing moth larvae feed on the weightless matter, as do cask moth larvae, which spend their entire larval lives within a cocoon they carry around—woven from the same gray material they feed on.

A food chain of arthropods is built upon those primary foragers. Larger mites and silverfish prey upon the smaller forms. The top predator in the dust is the pseudoscorpion, which preys on all the other inhabitants below it on the chain.

Household pseudoscorpions can grow to several millimeters in length. They look like the front half of a true scorpion, lacking the recurved stinger. They are the most mobile of the dust creatures because of their propensity to hitchhike. They wait like ticks in the weeds and then reach out and clamp down with one claw onto anything that moves by. They are found passively riding on all manner of other, larger arthropods, as well as on vertebrates.

If a pseudoscorpion in the house latches onto a house fly, it may eventually be carried out the window. If it snags a bird outside, it may be transported to the nest. Thirty different species of pseudoscorpions have been found in birds' nest habitats.

Drifting back the other way, microscopic bits of space dust filter in through the window, after having fallen down through the atmosphere from higher up. The organisms that live in house dust cannot eat that space dust. Graphene flakes and silicon carbide crystals are indigestible. They remain inert, blending with the rest of the dust. They will eventually be swept outside unchanged, where they will sink through the ground over the millennia and separate from the biosphere, on their way to joining the mineral basement strata of the planet.

House dust, on the other hand, remains in the biosphere, recycled through the living world by the creatures that can live by consuming it. Those life forms transform the dry matter they find into the substance of their own cells. This is a chemistry and a level of organization of a much higher order than anything that exists in the dust lanes that divide the stars.

PLANET EARTH EMERGED in space, from stardust. Then different kinds of dust emerged on planet Earth. These processes fostered the creation of our world and the wild diversity of life in its every corner.

We can easily imagine that this evolution could have repeated itself in many different star systems across the universe. What we cannot easily imagine is the transition point—from the carbonaceous muck in cracks on sterile rock, to the vastly more complex, proliferating cellular forms that emerged from that primordial soup. We just don't know how facile the first step—from the carbonaceous dust, to the first life forms—may be.

How many paths lead to the genesis of life? Are there many ways that self-replicating organisms could arise from the carbon sludge on a young planet? Is the appearance of life forms on barren worlds inevitable? Or is the intricacy of life a once-in-a-gazillion quirk that chanced to happen only here? The vastness of the empty space between the stars may preclude our ever discovering that answer.

But we do believe that of all the wonders we can see or imagine among the stars, the most miraculous complexity and potential to be found anywhere in the galaxy is floating right next to us—weightless in a shaft of morning sunlight.

NOTES

The nebula from which the sun formed was made of gas and dust. Most of that condensed of its own gravity to give birth to our star. The leftovers remained in orbit around the young sun. More dust was formed by condensation of atomic particles blown away from the sun's surface. The solar wind drove those extra dusts beyond the frost line. Beyond that line, the heat of the sun is so diminished by distance that water molecules bind together and form ice. The elements in solid space dust reflect the elemental makeup of the sun. They include carbon (in the form of graphite or silicon carbide—Williams & Cecchi-Pestellini, 2016) and oxides of aluminum and magnesium. The elements in house dust are very different, reflecting the makeup of animal skin. They include carbon and nitrogen (in organic forms) as well as sulfur and phosphorus.

Humans shed several grams of skin-derived matter per day. That can build up, in out-of-the-way places, into house dust. House dust sustains an ecology (Van Bronswiki, 1979) based on pyroglyphid mites and fungi, which are the primary consumers of dust. Other residents of this habitat, including the larvae of clothing moths and dermestid beetles, consume the fiber components of dust, such as hair. These creatures are specialized to live in a xerophytic environment, where they must either harvest the water they need from dry air (air at 30 percent humidity is less than 1 percent vapor-phase water at room temperature) or generate it from their own metabolism (e.g., the metabolic oxidation of a microgram of dry carbohydrate generates about half a microgram of water).

The household environment supports a wide variety of arthropods, many of which are residents (Bertone et al., 2016), while others are just passing through. Those organisms on the bottom of the dust-habitat pyramid are preyed upon by larger mites and silverfish; the pseudoscorpions (e.g., *Chelifer cancroides*, the house pseudoscorpion) occupy the top of the pyramid. Worldwide, there are three thousand species of pseudoscorpions (www.museum.wa.gov.au /catalogues/pseudoscorpions). Their hitchhiking (phoresy) is selective in its choices of living conveyance, defining the niches occupied by each species (Poinar et al., 1998).

REFERENCES

Bertone, M., et al. 2016. Arthropods of the great indoors: Characterizing diversity inside of urban and suburban homes. *PeerJ* 4:e1582.

Poiner, G. O., et al. 1998. Arthropod phoresy involving pseudoscorpions in the past and present. *Acta Arachnology* 47, 79–96.

Van Bronswiki, J. E. M. H. 1979. House dust as an ecosystem. *Recent Advances in Acarology* 11, 167–72.

Williams, D. A., & C. Cecchi-Pestellini. 2016. *The Chemistry of Cosmic Dust.* CP1 Group, Croydon, UK.

— 22 —

Red Planet

THE PRE-SOLAR NEBULA was a cloud of gas and dust containing all of the elements we now find in the planets, including iron. Planet Mars condensed from this nebula more than four billion years ago. Back then, when the planet was first finding its orbit around the young sun, it was already destined to become a red planet, because of the iron it incorporated into its sphere.

Early on, however, Mars was a blue planet, like Earth. The blue phase did not last. Seas evaporated, the atmosphere thinned, and the entire surface grew exposed to the environment of deep space. Iron atoms in the dust were oxidized to their red hematite form by ultraviolet light from the sun. Those rays have rained down upon the Martian desert surface every day since, up through the present.

Larger amounts of iron from the solar nebula were incorporated into the newly forming planet Earth. The stage was set for an Earth as red as Mars is. Indeed, there once was a place on Earth as red as the plains of Mars—red as the iron that colors blood.

But the evolution of the Earth diverged from that of Mars. Earth retained a thicker atmosphere that blocked ultraviolet radiation. So earthen iron became incorporated into the rocks still in its uncolored form.

Its burial continued to protect that iron from oxidation a few billion years later, when a transformative event swept beneath the clouds. Earth's atmosphere was thickened with oxygen—which was generated by photo-synthetic organisms that pumped it into the sky.

After another few billion years, a wide mountain range rose on Earth, over the young continent of North America. Those mountains would soar to ten thousand feet, only to be later disintegrated into dust, and spread flat across a wide plain. Over the millennia, that dust would be plowed by churning rivers in one season, then left to dry out in the next. This weathering pro-cess broke the iron out of the rocks, and exposed it to the oxygen in the air.

THIS MOUNTAIN RANGE began its decline three hundred million years ago, when Earth had an atmosphere of almost 25 percent oxygen—the most this planet has ever known. That was in the age of the great forests of the Carboniferous Epoch, when North America was a tropical continent riding on the equator. Dense plant cover pumped oxygen into the sky, and the fallen trunks of scale-covered trees piled up on the ground. The shallow Panthalassic Ocean covered much of what would become western North America. The western shore there was located where the Rocky Mountains are now.

Behind that prehistoric shoreline the Uncompahgria Mountains reached toward the stratosphere to intercept the winds. The ancient cordillera pre-cipitated rain along its highest ridges. Thundershowers incised canyons into the long, tall slopes. When the mountain peaks reached their greatest height and then ceased their growth, erosion was no longer balanced by ascension. Then the range began its decline.

Over the next fifty million years, North America itself began to rise. It shifted in latitude, rising away from the equator toward the pole. All of the continents of the Earth were then moving together, rotating in a great gyre. In the southern hemisphere, that rotation was pulling ancient Gondwana-land down toward the South Pole. The wide land mass grew ice-covered, and the Earth began to cool. The tropical carboniferous forests receded into their last refugia and then on into extinction. Earth's expanding south polar ice cap pulled megatons of water from the ocean onto the land, tipping the planet into one of its deepest ice ages.

Sea level fell three hundred feet. The beach beside the foothills of the Uncampahgria Mountains fled to the west and disappeared over the horizon. Flat continental-shelf seafloor was exposed to the sky. Cataracts through the mountain canyons slowed and flattened into shallow meanders where they emptied onto the new floodplain. Rocks carried downslope in flash floods came to rest in broad alluvial fans choking the mouths of canyons. Silt carried farther settled out in the riverbeds on the flatland.

THE UNCOMPAHGRIAS WERE great granite and sandstone monoliths. Granite is a tight matrix of quartz, feldspar, and mica crystals, each of them individually big enough to see by eye. Iron from the early Earth was locked up in the transparent, colorless flakes of mica.

When granite mountains are brought down by erosion, the hard quartz survives—it breaks down into grains of sand that last forever. But the mica is destroyed by the weathering process, reduced to kaolin clay. The iron dissolves out of the clay matrix into (oxygenated) rainwater. The transparent solution oxidizes and grows reddish. Hematite is produced, and it adheres to the grains of white quartz sand around it, tinting them red as well.

When the slopes of the eroding Uncampahgrias grew shallower, their destruction slowed. Precipitation declined as northward continental drift pulled North America into the horse latitudes. The canyons became mostly dry, but their streams were still swollen now and again by occasional thunderstorm-spawned flash floods.

The rare deluges pushed the rivers in the lowlands over their banks to writhe back and forth across the plains. Thick layers of sediment were transported away for hundreds of miles. Hills were flattened and gullies filled—exposing buried material to the sky and then burying it again—leveling the plains.

Ten million years of erosion reduced the entire volume of the Uncampahgrias to ground level. The mountain range was converted into a flat layer a quarter million square miles in area. The silts and sands covering that plain were the red color of hematite. For an instant in geologic time, the whole surface of the nascent southwestern US was covered in red mudstone, compacting below into a deposit of ruddy shale hundreds of feet thick. The vista from far above would have showed a patch of Martian red on a young continent drifting across Earth's marine-blue disk.

That interlude passed when the red plains were covered over. Northward drift of the continent into a milder climate zone, along with periods of ice-age dryness, converted much of western North America to desert. White crystalline sand blown down from Canada concealed the red southwest under shifting dunes hundreds of feet thick.

Those white sandstone deposits were then swamped by the returning ocean after the end of the Permian epoch, 250 million years ago. Over the millennia, the floor of that ocean was buried under the calcareous skeletons of countless settling diatoms. The white sandstone stratum was capped by layers of marine limestone.

PLANET MARS HAS a rocky shell that is uniformly solid. Its globe is covered in a single tectonic plate. The exposed Martian rocks contain almost no free water. They are locked in place. In contrast, the rocks in the surface layers of Earth are wetter. This may explain why Earth's lithosphere is so much more mobile.

Earth has a surface divided into a score of tectonic plates. Energies stirring the magma ocean on which they ride keep these plates drifting in different directions. They slide past one another along fault lines, or they seize up at their boundaries and eventually ride over and dive under each other.

Such plate tectonic collisions had consequences for the new southwestern American desert. Beyond its far western horizon, mountains were rising from the ocean. They enlarged in perspective as they encroached upon the shore. Those mountain terranes were carried on tectonic plates that were moving east and then subducting under the North American continent. The island arcs they brought were welded onto the shoreline, expanding the continental margin to the west.

The subducted plates (without their terrane cargo) continued drifting east below the surface of North America. When they reached the longitude of the desert southwest, they shouldered the plate above them upward. The layers of sedimentary deposits from the Permian Period, including the buried red shale, were raised a mile above their original elevations.

The uplift became the Colorado Plateau. The rivers flowing across this bench found themselves cascading with increased erosive power where they fell over the Mogollon Rim at the plateau's edge.

WE CAN SEE the ruddy color of Mars when the red planet drifts before us, against the star clouds of a distant evening. We can see the same color on Earth in places where the red sedimentary layer in the American southwest has been uplifted and exposed by erosion.

The uplift of the eastern reaches of the Colorado Plateau elevated the San Juan River watershed a thousand feet above the river's terminus. As a slow-flowing watercourse, the river had not shown great erosive potential. But accelerating through a thousand-foot drop from its elevated section, the currents grew stronger. Over millions of years, they deepened the river's canyon, cutting down through the uplifted rocks.

The river captured smaller creeks that were carving their own gorges as they ran off the rise. When the streams had cut their beds back down to their pre-uplift levels, the water flattened out and grew calm again. Over time, the drainage undercut the surrounding walls, and widened its network of canyons into broad shallow valleys, which eventually merged. All that remained of the uplifted strata were isolated sections of free-standing thousand-foot walls.

These pillars now stand in Monument Valley. They reveal a vertical cross section of the sedimentary layers just beneath the top of the Colorado Plateau. Their sheer-sided buttes and mesas are carved from the white cliff-forming sandstone of the Permian period. Below them, at ground level, the bases of these monuments stand on skirts of fallen rock as red as a Martian desert. That lower layer is the red shale derived from the ancient Uncompahgria Mountains. It now lies mostly hidden, buried six hundred feet below the top level of the plateau. But it is exposed at the bases of the monoliths there in the valley.

THE OAK CREEK Fault has cut south across the Colorado Plateau for tens of millions of years. The block of crust on the east side of the fault has subsided, elevating the block on the west. The rocks between the blocks have been ground to bits by the moving fault. When runoff spilled along the fault line, the pulverized sand was washed out and a creek bed became established in the notch. Where it continues over the side of the Mogollon Escarpment, Oak Creek falls across a three-thousand-foot drop at the edge of the Colorado Plateau.

The high plateau is snow-covered in the winter. Spring runoff through the creek now falls with enough hydrodynamic force to occasionally carry boulders. A recently deposited layer of volcanic basalt on the eastern side of the creek was undercut by erosion and broken by the tumbling cascade. Basalt fragments carried downslope increase the erosive force of Oak Creek.

The stream has deepened its canyon to the south and cut its headwaters notch back across the plateau to the north. In the northern extension of the canyon the limestone bedrock of the top of the plateau stands exposed in cross section. Further down, cliff-forming sandstone appears. Balanced boulders of hard cap-rock protect the tops of sheer pedestals. Between them, the upper levels of the plateau have fallen away into the flanks of the escarpment.

Farther down, the creek exposes the lower levels of the sandstone layer, and then the red rocks underneath. Red walls guide the creek down the escarpment through miles of its descent. At the base of the slope, those cliffs are eight hundred feet tall.

Oak Creek Canyon is lined by ruddy walls all the way down its length to the plains below. The conifer forest at the top of the creek grades into mesquite and ocotillo along the descent. The landscape finally flattens into saguaro cactus habitat in the lowlands far to the south. Buttes and mesas stand free beyond the base of the canyon. The red shale formations tint the monuments in the colors of a windswept Martian desert landscape.

EIGHTY MILES NORTH of the headwaters of Oak Creek, there lies another canyon. The watercourse flowing through this one has a gradient drop of over a mile from its head to its terminus. It has cut a gorge a mile deep through the plateau that has risen across its path.

In this grand canyon the Martian-red Permian shale layer appears as a sedimentary ribbon running five hundred feet below the rim. The layer is dwarfed by the strata beneath it, all of which were laid down earlier in time.

The layers in the walls of the Grand Canyon continue down almost five thousand feet below the red strata. The descent from the red Permian shale down to the narrow watercourse at the canyon bottom represents almost two billion years of Earth history. This is roughly half the distance, in geologic time, between the creation of the red desert on Earth, and the creation of the red deserts of Mars.

The early surface features of Earth have been destroyed by geological recycling. The oldest landscapes that survive here are recent compared to the ages of the older surface features of Mars. The hills of Noachis Terra on the Martian southern shield are 4.5 billion years old.

Across distances in space, we never come very close to Mars (about thirty-five million miles at the nearest). But we can come close to the experience of its ancient red landscapes with a much shorter trip, on a visit to the American southwest.

NOTES

The iron on the surface of Mars is oxidized to red by the solid-state liberation of oxygen by ionizing ultraviolet radiation coming down from space (Dauphas & Greenwood, 2010). On the surface of Earth, iron is oxidized by molecular oxygen generated through photosynthesis. The end result is the color of the exposed hematite in both cases.

The Carboniferous Period saw the genesis of coal from layers of plant matter that piled up on the ground and were eventually buried. (It may be that the wood-rotting fungi, which prevent such buildups of fallen carbon today, had not evolved three hundred million years ago.) At that time the Uncompahgria Mountains stood east of the location of the future Colorado Plateau (Blakey & Ranney, 2018)—prior to their complete destruction by erosion. The resulting red sedimentary layer (Hermit Shale) on the upper wall of the Grand Canyon is thinner than the red layers on the walls of Oak Creek Canyon. The red layers surrounding Oak Creek include the Schnebly Hill sandstones and the Supai formation (Ranney, 2015). This red sedimentary stratum is called the Organ Rock Shale in Monument Valley. Tall spindles (hoodoos) in Monument Valley and elsewhere in the American southwest remain standing where hard pieces of cap-rock protect the material below them from top-erosion.

REFERENCES

Dauphas, N. N., & R. C. Greenwood. 2010. Iron and oxygen isotope fractionation during iron ultraviolet photooxidation: Implications for early Earth and Mars. *Earth & Planetary Science Letters* 458, 179–91.

Ranney, W. 2015. *Sedona through Time: A Guide to Sedona's Geology.* (Self-published).

Blakey, R., & W. Ranney. 2018. *Ancient Landscapes of Western North America: A Geologic History with Paleogoegraphic Maps.* Springer: London.

— 23 —

Daphnis

DAPHNIS IS A grand, snowy-white mountain five miles tall. It rises from a plain that stretches out for tens of thousands of miles in every direction. The plain is made of icy rubble, but there is a rock-free buffer zone ten miles wide on either side of the mountain. The two sides of this cleared zone have parallel, straight edges; they form the borders of a chasm twenty miles wide and infinitely long. Its edges converge in perspective as they recede off toward the distant horizon. The crevasse has no bottom; it falls away into an infinite abyss—its depths as black as outer space. Daphnis sails through this channel like a huge iceberg.

Looking ten miles across the channel from the mountain, the individual pebbles on the edge of the plain are too small to see. The distant margin of their landscape looks like a sharp layer of fog. It shimmers under the chill night sky, translucent, the light of distant stars shining through it. The edge has a definite texture, and if you watched it closely you would see that it is moving to the left at one mile per hour. This is the speed at which the mountain is drifting to the right, past the distant shore.

Below the edge of the plain, there is only blackness. The plain is a floating, two-dimensional sheet of white bits of ice. Off to the right, its edge sharpens as it recedes into the distance. But to the left, the edge is roiled into billows of cloud that arch out above and below it. Those clouds have been whipped up by the passing mountain. Their shadows stretch out for miles across the flat expanse behind them.

If you turned to look in the opposite direction from the top of the mountain, the edge on the other shore would also be seen to be moving to the left. That means the plane on one side is sliding past in the opposite direction from the plane on the other side. In this opposite direction, the edge approaching from the right again sharpens as it reaches away, while the edge receding to the left is again consumed with billowing clouds. So the mountain appears to be drifting in opposite directions, depending on which way you are facing.

The wakes of cloud raised on the receding shores by the passing mountain are not stirred up by a pressure wave. Daphnis meets no resistance as it glides down this channel; it moves through a vacuum. The disruption of the plain is caused by the gravity of the mountain pulling on the ice particles across the gap. After the mountain has moved on, these clouds of particles settle back down to their own ground level, restoring the infinite horizontal surface.

The two edges of the Daphnis channel, across the rock-free zone, are nearly identical, but the vistas in the background far behind them are decidedly different from one side to the other. The view one way rises out into the starry night sky. But the opposite direction provides a broadside view into a world of storms. This backdrop looks up onto an imposing wall banded in yellow and brown clouds—the looming disk of Saturn. The planet is so close that the motions of the jet streams tracking across its atmosphere can be seen from Daphnis. The round face of Saturn is bisected by a set of sharp, dark, parallel lines: the shadows of its rings. The tawny light reflected from the clouds brightens the planetary side of the Daphnidian landscape, almost as brightly as the sun illuminates the other side.

Daphnis is a moon of Saturn. It is an irregular block of ice that orbits imbedded in Saturn's rings. Pebbles from the ring plane float toward Daphnis slowly enough that, were you standing on it, you could pluck them from the air as they approach. These rocks are made of ice, but at the temperature of space almost a billion miles from the closest star, they are as dry and hard as feldspar.

Daphnis spins on its axis once per orbit: it always shows the same face to Saturn. A steep ice ridge rises over its surface and runs completely around its circumference. The ridge is made of particles from the rings, which accumulate where the narrow ring plane intersects the floating mountain's equator. The moon is made from the ice that it sweeps up as it clears the gap through which it orbits. As it moves, it increases in mass.

The plane of ice particles Daphnis sails through has no end: its curvature shrinks away across the sky and disappears behind Saturn, only to return on the other side—a circular sea of frozen water. Every particle in the ring is, like Daphnis, in orbit around Saturn. Bits of ice and larger moons follow similar orbits around the planet. They all move more slowly the farther away they are. So Daphnis moves faster than the ring material that orbits on its spaceward side, while the material in the ring on the planetary side of Daphnis moves faster than Daphnis itself. Thus, from the two perspectives of the rings on its opposite sides, the orbiting mountain appears to be moving in opposite directions.

AS DAPHNIS COASTS above the sunny side of Saturn, an arc of darkness approaches—steadily consuming the planet's leading edge. That sunset line reaches from pole to pole, advancing across the planet's disk from east to west as the night-side hemisphere rotates into view. As the moon moves above Saturn's shadowed side, the dayside face of the planet wanes to a crescent. The trailing, ice-bright rings behind Daphnis—the most reflective surface in all of the solar system—fluoresce, backlit by sunlight. As Daphnis slides farther over the night-side darkness, the planet's shadow moves down the length of the rings, eclipsing their brilliance.

Daphnidian sunset sees the thinnest of crescents backlit along Saturn's circumference. Atmospheric clouds on the curved horizon flare against the backdrop of blackness. The sharp yellow edge reaches from pole to pole, halfway around the planet's edge. After sunset, the glimmering edge fades into shadow while the evening star sinks toward the planet's darkened curve. From Saturn's perspective, Jupiter is the evening star. Its light is variable, dimming and brightening as Saturn moves through its seasonal cycles.

The night side of Saturn is bathed in a ghostly alpenglow. An equatorial band of ring-shine reflects the light from the sunlit limbs of the rings. They flank the darkened disk on either side. The amount of ring-shine varies with the Saturnian seasons. From the sun's perspective, the disk of the rings flattens with the approach of the equinox. At the minimum, they are edge-on to the sun. At that time, they disappear into a single thin line. Almost no sunlight is reflected from the rings then, and the dark side receives minimal ring-shine. So the back of the planet fades to a black, a star-free disk imposed on the infinite stellar backdrop of space.

Cumulous silhouettes in Saturn's night-side cloud deck flash into view now and then. Their contours are backlit against the blackness by strokes of silent lightning. Above and below the darkened tableau, the planet's poles are crowned in shimmering pink aurorae.

JUPITER HAS FOUR moons about the size of Earth's moon. The three outermost of them are covered with shells of ice fifty miles thick. The ice floats on deep spherical oceans, above their moon's solid cores. Saturn once had moons like that, but their orbits were unstable. One after the other, the ancient spheres spiraled downward to impact the planet. Only one, Titan, now survives in a stable orbit.

The last of Saturn's major moons to be lost was Veritas. It gradually descended toward Saturn until its shape became distorted. The planet's gravity was raising tides in the solid ice shell around the moon, stretching its shape.

The oblong distortion increased the closer Veritas approached. When the tidal forces exceeded the structural strength of the ice, fissures grew across the moon's cratered surface. With each pass, those fissures grew longer and more branched. Where these cracks merged, the ice fractured. Smaller blocks were isolated by fault-scarps on all sides. They separated from the surface and were lifted up and away by Saturn's gravity.

When those blocks rose above their moonscape, the ocean below them was released to follow them into space. The larger pieces of the moon's shell jostled and settled, pushing into the waters underneath, which were displaced skyward in towering geysers. The moon disappeared in a cloud of sparkling snow as its waters exploded weightless above the surface and froze in the chill of space.

The larger blocks of the moon's ice shell lost the support of the ocean upon which they once floated. They fell against the solid rocky core to shatter further. A mass of ice fragments expanded away through space in all directions.

The round core, dark and cracked, emerged from the mass of floating, tumbling debris with its descending momentum still on course. Its path bent evermore downward toward fiery destruction on impact with the planet below. An armada of smaller fragments followed it into the atmosphere. They trailed parallel steaming contrails as their courses slanted through the clouds. Veritas would disappear from the skies of Saturn forever, but it would leave a brilliant legacy.

Countless pieces of rubble that had spread away from the disintegrating moon found their descent bent back upward into stable orbits. They would provide the material that would become Saturn's rings. The frozen shards would be reduced to the size of gravel and spread farther apart by millennia of collisions with each other.

Those particles had once been a spherical ocean, and a thick shell of ice floating upon it—covering Veritas. Veritas was once a frozen world a thousand miles in diameter. But it lost one of its dimensions and expanded the other two. It became a flat circular plane of icy debris fifteen billion square miles in area, adding a spectacular set of rings to a previously ordinary planet. A miniscule amount of that sparkling dust eventually consolidated to produce Daphnis.

SATURN'S RINGS ARE not the featureless flat planes first described by the early astronomers. They have been found to be grooved and gapped as densely as a vinyl record. One of the smaller gaps is cleared by the orbiting passage of Daphnis. A larger gap is cleared by another moon, Pan—a floating icy mountain ten times more massive than Daphnis. Pan clears the Enke Gap in the rings, a channel ten times as wide as the gap cleared by Daphnis. The ridge of ring material deposited on the Pandean equator is taller and wider than that on Daphnis, leaving Pan looking like a rounded ravioli.

The largest gap in the rings is cleared not by a moon moving through it, but by a moon thousands of miles above it. This is the ice moon Mimas, a body ten thousand times as massive as Pan. The spherical form of Mimas is distorted by tidal bulges, just as Veritas was distorted when it first approached Saturn. But Mimas is not on course for destruction. It has settled into a stable orbit and is descending no further.

Mimas has cleared a gap in the rings a thousand miles wide, thirty-five thousand miles below its own orbital distance above Saturn. The moon reached across the rings to clear this gap using its own gravity. Mimas evicted ring particles that were in resonance with it: their orbital periods were exactly twice that of the distant moon. The gap that was created is called the Cassini Division. It is an old and permanent feature of the rings, as well established as Mimas is in its orbit.

The Cassini Division is populated only with stray particles that chance to wander in from elsewhere in the rings. Should any of those come to settle in circular orbits there, they will be ejected from their paths by the invisible hand of Mimas, as were the particles before them.

THE MORE THE rings of Saturn are understood, the more it appears that they are a microcosmic model for the entire solar system. The solar system is also a disk of round planets and smaller rubble, orbiting a gravitational center provided by the sun. Just as Mimas reaches across space to maintain a distant gap in Saturn's rings, so has Neptune's gravitational influence reached into the broad ring of debris orbiting the sun beyond the orbit of Pluto. Neptune has cleared a gap in the Kuiper Belt, evicting particles with orbital periods in 2:1 resonance with its own period. Similarly, Jupiter alters the orbits of the debris in the plane between it and Mars, scattering away material once in orbit there, creating the gap now occupied only by the asteroids.

All orbiting bodies exert gravitational pulls on other bodies orbiting nearby. The inner Galilean Moons of Jupiter have pulled each other's orbital periods into a 4:2:1 resonance. Earth and Venus have reached across space to pull their orbits into a near-perfect 8:13 resonance.

When Uranus was discovered in 1781, its orbital period was deduced from its mass and its distance from the sun; but this calculated period did not match what was observed. In order to reconcile the true orbital period with the calculations, Urbain Le Verrier inferred the existence of another planet, another billion miles farther out from the sun, the gravity of which was perturbing the motion of Uranus. Using Newtonian mechanics, he calculated where that distant planet could be found based on the deviations in the orbit of Uranus. Astronomers following his directions discovered Neptune in 1846.

Elements of Saturn's rings are still being discovered, even today, the same way, by mathematical inference. Mark Showalter and others noticed a gravitational perturbation of the orbits of ring particles in the Enke Gap. Their calculations led them to discover Pan in 1990. Daphnis was discovered in 2005 through similar inference from its gravitational effects on the particles in the Keeler Gap.

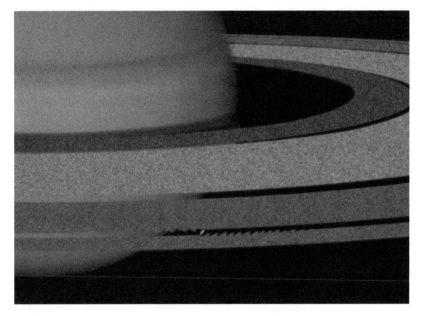

Other gaps in the plane of Saturn's rings are likely cleared by as yet undiscovered moonlets even smaller than Daphnis. Complex banding patterns that cover most of the rings likely originate as harmonic gravitational resonances between the particles and the moonlets. The sources of all these gravitational interactions remain to be described by future astronomers.

NOTES

In earlier days, Saturn held large moons, like those of Jupiter. Their orbits were unstable because they sailed through the disk of dust and debris left over from Saturn's own formation. The drag from this material stole the forward momentum that kept them in orbit. One hypothesis for the origin of Saturn's rings is that a Saturnian moon the size of Ganymede spiraled down to crash into the planet (Canup, 2010). Its miles-thick mantle of ice was shattered as the moon exceeded the Roche limit for tidal instability. Much of the shattered ice mantle of this moon scattered back into orbit to form the rings, as well as Saturn's smaller icy moons such as Mimas. Other measurements of Saturn's rings can be interpreted in terms of more recent ring origins, though we lack feasible explanations for the process that might drive such events (J. Cuzzi, cited in Voosen, 2017).

The farther they are from Saturn, the slower the bodies in Saturnian orbit travel. The velocity of orbiting bodies at a particular distance from the planet is calculated by $(\sqrt{G \times M_S})/d$—the square root of G × the mass of Saturn, divided by the distance of that particular orbit to the center of the planet (G is the gravitational constant). Titan flies at 5 km/second, orbiting at a distance of twenty Saturn radii, completing an orbit in fifteen days. Daphnis, at a distance of one Saturn radius above the clouds, flies at 16.7 km/second, completing an orbit in fourteen hours. The various broken ice particles in the rings follow these same orbital dynamics. Analysis of data from the Cassini spacecraft's mission to Saturn shows the rings to be only ten meters thick (Tiscareno et al., 2007).

Venus and Earth are linked in a 13:8 orbital period resonance; the ratio is not perfect, indicating that the resonance is still evolving. Just as Venus, the evening star as seen from Earth, never moves farther than 47° from the sun, so Jupiter remains within 30° of the sun from the perspective of Saturn. From the perspective of the sun, Saturn's rings shrink to a thin hairline twice per Saturn orbit (on the Saturnian equinoxes) once every fourteen and a half years. On its dark side, the Saturnian aurora is visible; it is pink, colored by the fluorescence of hydrogen (Dyudina et al., 2016).

Daphnis orbits in the Keeler Gap. Other moons clear other gaps: Pan orbits in the Enke Gap. Atlas rides on the outer edge of the A ring. The Cassini Division between the inner B ring and the A ring was not created by a co-orbital moon but by a 2:1 resonance between the orbital periods of the ring particles and Mimas, a moon distant from the division. The Cassini Division and even the Enke Gap are visible by telescope from Earth. Giovanni Cassini discovered the gap named after him in 1675.

The gravity of every particle in the rings pulls on every other particle there. Most of those attractions are minuscule, especially if they come from tiny chips of ice and are diminished by great distances. Nonetheless, some of these individual gravitational interactions have visible effects. The gravity of Mimas reaches across the rings and perturbs the motions of a set of ring particles that orbits Saturn with a period exactly twice that of Mimas. When they pass below that moon on their closest approach to it, each of those particles feels the maximal gravitational attraction from the Mimas. And they feel that at the same point in their orbit, on every cycle. They are in "resonance" with Mimas.

The pull on the particles orbiting exactly twice as often as Mimas is always maximal in the same compass direction. Over millions of years, that pull moves the resonant particles up toward Mimas at the point of their closest approach. The concerted attraction warps their circular orbits into ellipses. This removes those particles from their earlier circular orbits. They find themselves on new courses, traveling at nonresonant speeds. This clears a gap in the rings almost three thousand miles wide, from which the particles have all been evicted.

REFERENCES

Canup, R. M. 2010. Origin of Saturn's rings and minor moons by mass removal from a lost Titan-sized satellite. *Nature* 468, 943–46.

Dyudina, U. A., et al. 2016. Saturn's aurora observed by the Cassini camera at visible wavelengths. *Icarus* 263, 32–43.

Tiscareno, M. S., et al. 2007. Cassini imaging of Saturn's rings 11: A wavelet technique for analysis of density waves. *Icarus* 189, 14–34.

Voosen, P. 2017. Cassini observations show Saturn rings are a recent addition to the solar system. *Science* 358, 1513–14.

— 24 —

Photon

THE PHOTON AWOKE with a bang. It had spent the previous ten billion years as part of a hydrogen atom. Inside of a young star, the atom had been stripped of its electron. It was left as a bare proton, surrounded in the star's core by other protons. They were compressed to densities far greater than metal, and boiling at white-hot temperatures in the millions of degrees. There, half a million miles below the star's surface, the proton crashed head-on into a proton/neutron pair. The collision fused the two particles into a helium nucleus. As a byproduct, the fusion produced the photon—which was born flying at the speed of light.

The newly born light wave traveled less than a picometer before it collided with another proton. The photon's energy was there converted into matter—one electron and one positron, particles with equal masses, but opposite charges. The two particles instantly destroyed each other in an antimatter annihilation, recreating the photon, which took off again in a different direction. It was then absorbed by an electron, which recoiled in the opposite direction and re-emitted the photon at a lower energy. The photon repeated interactions like these millions of times in the first second of its existence. Even though it always travelled at light speed, its course was a tight clump of zigs and zags. Its net movement was only inches per hour.

A MILLION YEARS after its creation, the photon's random path had taken it to the uppermost layer of its star. Its switchback course had finally spanned the star's radius. In air, it would have covered that linear distance in a few seconds. By now, its total travel distance was equivalent to halfway to the nearest galaxy.

The photon had been born as a high-energy gamma ray. But over the course of its travels, it had given away most of its energy to its surroundings as heat. Its own energy was now only that of a photon of visible light. It would be perceived as yellow. The number of collisions it suffered per second had decreased as the density of matter in the star thinned out toward the surface.

Then one day the continuous succession of collisions ceased. The photon went from flying through tiny distances between a million course changes every second to flying in a perfectly straight, uninterrupted line. The previous second, it had covered only a micrometer in its zigzag path through the white-out stellar brightness. But in this new second, it flew through the cold blackness of space for two hundred thousand miles. The star in which it had spent the last million years was a shrinking ball of fire falling away behind it.

Even though the photon was traveling at the speed of light that day, as always, the starry backdrop around it was suddenly static. Like a moonless night-sky mountain vista frozen in time, nothing changed—from one hour, or day, or year to the next. Time appeared to have been suspended. The constellations were the same as the ones seen on Earth. The stars did not twinkle, nor did they track across the frame. They were fixed in place, like diamond-dust glitter on a black wall.

Most photons travel forever, centered in an unchanging void. The constellations that surround them take hundreds of years to show any change. After thousands of years the stars grow sparser as the photons move out of the galaxy and into the dark intergalactic spaces. Faintly luminescent smudges may drift past over a span of a million years or so. But aside from those occasional far away galaxies, light speed itself reduces to nothing more than eternal interment in motionless blackness.

A FEW DAYS after this photon's release, its birth star had shrunken behind it from a disk to a point source. It was the brightest star in the sky; its magnitude was unchanging. It was positioned adjacent to the star patterns of

Orion and fixed at the head of the blue-white stars of Canis Major: the star of the photon's origin was Sirius.

Over the course of this flight, there would be but one change in the vista in which the photon rested. One of the stars in the stellar backdrop near Vega would brighten slightly. None of the other stars changed, but this one grew a distinctive yellow brilliance.

After eight years, Sirius had faded to magnitude minus two, while the yellow star had grown to be the brightest light in the sky. It was the center of a thin disk of debris the photon was approaching. Absorption of the photon by matter in that disk would end its flight.

The photon was on course to pass through this debris disk on its way out of the galaxy. It flew toward the disk center, passing a large dark spherical body in the outskirts of the system, and approached closer to a similar planet an hour later. Other large, cold bodies also orbited the yellow star, but they were so far off that they appeared as stars themselves. Still, these were moving stars, sliding past the deeper backdrop.

The star at the center of this system was resolving into a bright disk, and it too was moving—slowly, to the left. The photon was on course to pass it by.

A few hours later, another point of light resolved into another dark disk, to the right of the central star. This body was lit along one edge with a crescent of yellow, reflected from that star. In contrast to the larger bodies farther out, the rocky surface of this planet was visible through its transparent atmosphere. The body did not drift to either side, it only grew larger.

Photons from the central star in this system rained down on the day side of the approaching planet. They were absorbed by electrons in molecules on the surface, breaking the bonds between atoms, as they do in other planetary systems. But on this planet, the energy of incoming photons was used to break the bonds in H_2O molecules, liberating the oxygen as a gas into the planet's atmosphere. In this regard, the planet was different from every other planet in the known galaxy.

As the photon approached the planet, the vista ahead came to center on a spot on the upper edge of the disk, in the darkness just below the bright margin of the crescent terminator.

The photon had been traveling as a member of a convoy of photons that had originated from points all across the million-mile diameter of Sirius. They had all chanced to head off in the same direction, at the same time.

They travelled together in space unseen—they were only visible from the direction straight down their flight path. Any of them that were detected, for example, absorbed by a molecule, would not complete the flight.

FROM EARTH, TO look at Sirius is to look down a converging stream of photons. They fly in every direction from the star's surface. Earth's disk eight light years away makes up a tiny, tiny fraction of the sky as seen from Sirius. Yet the star puts out so much energy in all directions that the fraction of its output that reaches us is still enough to bathe our entire planet constantly in Sirius-light. The star has shone steadfast in earthly skies for every pair of eyes anywhere on the globe that has chanced to look up on a clear evening over the past million years.

In the last millisecond of their group flight, the approaching convoy of photons containing our yellow one passed through Earth's atmosphere. Theirs was a rainbow flight: they carried a range of energies. They included photons from the infrared to the ultraviolet and all the visible colors in between. Their speeds through the vacuum of space had all been the same, but when they entered the medium of air they slowed a little—some more than others. Photons with higher energies, toward the violet end of the spectrum, were slowed more than those toward the red. The effect was magnified because these photons entered the atmosphere at an oblique angle—from a direction an observer would see closer to the horizon. So they passed through a maximal length of the atmosphere.

The light of a rising Sirius passed through pockets of air of different densities—different temperatures, different stratospheric wind speeds—and the starlight came to life. It was refracted—the white light of the mixture of photons was separated into bands of individual, prismatic colors. The constant cold white mix flying in space was changed for Earth-bound observers into a twinkling focal point that occasionally flickered with scintillae of prismatic red or blue.

In the last picosecond of its flight, the yellow photon and others flying with it entered the eye of a creature on the surface of the Earth, who chanced to be watching Sirius rise. The photons' paths were diffracted by the transparent radial structures of the lens of that eye, producing the radial spikes that appeared to the viewer around the edges of the image of the star.

The yellow photon was absorbed by an electron in a pigment molecule on the observer's retina. The absorbed energy caused the electron to recoil, breaking the bond it was part of. The electron released a set of low-energy infrared photons as it settled back down into its bonding orbit.

During the few picoseconds that the bond was broken, a signal was generated on that retina for transmission to the observer's brain. There, an image was forming, of Sirius sparkling through the twilit dusk, in the light of photons that had been flying, since their birth, for a million years.

NOTES

The relationship between matter and energy is described by the equation $E = mc^2$, where the value of $c^2 = 3.8$ sextillion (miles/hour)^2. The large value of that constant suggests that tremendous amounts of energy are packaged into small masses during the interconversion of energy and mass. This tale begins with the fusion of a proton with a deuteron (a hydrogen$_2$ nucleus) in the core of a star. In that reaction, a very small amount of the mass of the starting particles is converted to energy (produced as a high-energy gamma ray photon) during the production of a helium$_3$ nucleus. The radiation liberated by the stellar fusion of those atomic nuclei provides the energy that heats stars and resists their gravitational implosion. Stars exist in an equilibrium between explosive disruption and gravitational collapse. Gamma rays produced in stellar fusion reactions give up most of their energy to the star as heat as they bounce around inside it over hundreds of thousands of years (Mitalas & Sills, 1992) before they finally escape from the stellar surface as lower-energy visible-light photons. The eye can sense individual photons (though multiple photons must be sensed together before a signal is generated in the optic nerve). Photons are always moving at the speed of light. When they encounter a molecule, photons can be reflected or absorbed by the electron cloud that holds the molecule together. Electrons that absorb energy are raised to higher orbitals in the cloud. This can break the bonds between atoms in the molecule. On the unusual planet where this tale ends, the absorption of photons on the surface is sometimes used to break the bonds between atoms in water molecules to liberate oxygen. On most planets, this does not happen, but on this planet a special chemistry on the surface focuses photon energy. That is the chemistry of photosynthesis (Joliot & Kok, 1975).

This story is an expansion of one sentence from a previous treatment (Daubert, 2009) of this topic.

REFERENCES

Daubert, S. 2009. The light fantastic. In *The Shark and the Jellyfish*, 165–70. Vanderbilt University Press, Nashville, TN.

Joliot, P., & B. Kok. 1975. Oxygen evolution in photosynthesis. In Govindjee, ed. *Bioenergetics of Photosynthesis*, 387–412. Academic Press, New York.

Mitalas, R., & K. R. Sills. 1992. On the photon diffusion time scale for the sun. *Astrophysical Journal* 401, 759–76.

[INDEX]